STATISTICAL METHODS IN BIOLOGY

Third edition

Generations of biologists have relied upon this useful book, which presents the basic concepts of statistics lucidly and convincingly. It recognises that students must be aware of when to use the standard techniques and how to apply the results they obtain. The reasoning behind the more important procedures is carefully explained. Since many biologists do not have a strong mathematical background, the arguments are gauged in terms which can be easily understood by those with only an elementary knowledge of algebra. Unlike many other introductory books, mathematical derivations are avoided and formulae are only used as a convenient shorthand. Although the subject is presented with great simplicity, the coverage is wide and will satisfy the needs of those working in many disciplines.

New material for this third edition includes a discussion of the role of pocket calculators. These should have at least automatic calculation of reciprocals, squares, square roots, natural logs and exponentials. It is also recommended to have built-in statistical facilities for means, standard deviations, correlation coefficients and simple regression. For regression with three or more independent variables there should be built-in solving of linear equations and/or simple matrix algebra (including inversion).

This covers all the statistical methods presented in the book. However, many readers will want to take advantage of their acquaintance with at least elementary computer facilities. This is especially desirable in manipulating large data sets, but great care must be taken in choosing appropriate statistical software. A special chapter is devoted to a discussion of problems associated with numerical calculation, electronic calculators and computers.

STATISTICAL METHODS IN BIOLOGY

Third Edition

Norman T. J. Bailey, M.A., D.Sc.

PUBLISHED BY THE PRESS SYNDICATE OF THE UNIVERSITY OF CAMBRIDGE
The Pitt Building, Trumpington Street, Cambridge, United Kingdom

CAMBRIDGE UNIVERSITY PRESS
The Edinburgh Building, Cambridge CB2 2RU, UK http://www.cup.cam.ac.uk
40 West 20th Street, New York, NY 10011–4211, USA http://www.cup.org
10 Stamford Road, Oakleigh, Melbourne 3166, Australia
Ruiz de Alarcón 13, 28014 Madrid, Spain

First published by Edward Arnold 1959
Second edition 1981
First published by Cambridge University Press 1991
Third edition 1995
Reprinted 1996, 1997, 2000

Printed in the United Kingdom at the University Press, Cambridge

Typeface Times 10/12 pt

A catalogue record for this book is available from the British Library

Library of Congress Cataloguing in Publication data
Bailey, Norman T. J.
Statistical methods in biology/Norman T. J. Bailey. – 3rd edn.
p. cm.
Includes bibliographical references (p.) and index.
ISBN 0 521 47032 3 (hc). – ISBN 0 521 46983 X (pb)
1. Biometry. I. Title.
QH323.5.B33 1995
574′.01′5195–dc20 94-4693 CIP

ISBN 0 521 47032 3 hardback
ISBN 0 521 46983 X paperback

Contents

Preface to the third edition

This book was originally planned to meet a continuing demand for an elementary text to provide workers in the biological and medical sciences with the basic statistical techniques required in practice.

Accordingly, the mathematical symbolism used is the bare minimum required for a concise description of the essential types of analysis. The reader need have no more than an elementary knowledge of algebra. No trigonometry, geometry or calculus is used. In most cases the main text gives a fully worked numerical example of each different type of analysis, in addition to the general discussion and summarising formula.

Naturally, there has been great progress in statistical science over the years. Much of this has involved theoretical modifications and developments, but there have also been implications for elementary applications. Rather than attempt to cover a multiplicity of new refinements it was decided to concentrate on essentials. The content of the second edition has accordingly been carefully revised, updated and corrected where necessary.

It should be emphasised, however, that it is now some 13 years since the second edition appeared. This was at a time when pocket calculators were becoming widely available and having considerable influence on the speed and efficiency with which elementary statistical analyses could be carried out. However, at the same time there were still many countries where the older desk calculating machines continued to be used. It was, therefore, considered advisable to continue including a good deal of material relating to the operation of such machines – especially with regard to checking for errors and using short-cuts. It is important to realise that this situation has now gone for ever, thanks to the rapid spread, during the past decade, of advanced pocket electronic calculators with built-in statistical facilities, and of personal computers, including small but powerful lap-top versions, that can handle large data sets and employ specialised statistical software. The new third edition has therefore been extensively revised to reflect these changes.

An important feature is the 'Summary of statistical formulae'. This is intended to be used for day-to-day reference by the worker who already has some acquaintance with statistical ideas and who merely requires to have his or her memory refreshed. It is important that the formulae in the summary should not be applied blindly and automatically without proper regard for their suitability. If there is any doubt about the correct way to handle any particular statistical situation, advice should be sought from a qualified biostatistician. A selection of the most frequently used statistical tables is also provided in the appendix, so as to make the book reasonably self-contained for the purpose of everyday use.

ACKNOWLEDGEMENTS

Professor Sir Ronald A. Fisher, Dr. Frank Yates and Oliver and Boyd kindly allowed me to abridge Tables I, III, IV and VI from their book *Statistical Tables for Biological, Agricultural and Medical Research*. Some additional material has been incorporated from Tables 12 and 18 of *Biometrika Tables for Statisticians*, Vol. I with the permission of the *Biometrika* Trustees. I am grateful to Professor John W. Tukey and to Ciba–Geigy for permission to reproduce the table for the Wilcoxon test for pair differences taken from their *Scientific Tables* (7th edn.). In addition, the table for Kendall's rank correlation coefficient is reproduced by permission of the publishers Charles Griffin and Company of London and High Wycombe, adapted from Kendall *Rank Correlation Methods* (4th edn.), 1970. A simplified version is in fact provided, giving only critical values for certain convenient levels of significance.

Finally, I should like to thank the present publishers, namely the Cambridge University Press, for giving me the opportunity to revise the book yet again in a form that will, I hope, retain its character as a reasonably comprehensive, but elementary guide, at the same time maintaining the price at a level which is still relatively low by current standards.

Lauenen, Switzerland *Norman T. J. Bailey*
June 1994

1

Introduction

The subject-matter of the biological and medical sciences is remarkable for its richness and complexity. Moreover, the wide range of variation observed in both organisms and their environments is frequently analysable into simpler components only with great difficulty. Suppose we want to compare the behaviour of two different animal populations. Not only does each population consist of individuals differing amongst themselves with regard to factors like sex, age, physical measurements, coloration, susceptibility to disease, aggressiveness, etc., but the patterns of behaviour in which we are interested may themselves be fairly complicated. For these reasons, much biological work tends to be comparatively quantitative in nature. In the more exact sciences of physics and chemistry, on the other hand, we find that irreducible variation is usually fairly small, and often consists of little more than experimental errors. The latter can, as a rule, be virtually eliminated by averaging over several repeated determinations.

In biological sciences, therefore, inherent variation must be accepted as basic, and must be handled as such. This certainly makes numerical arguments more difficult. We may talk about the average number of eggs laid by a certain species of bird under particular environmental conditions or the proportion of subjects protected by an immunising vaccine. But these average figures conceal the fact that specific instances may easily show very different results. Some females will produce very large clutches, others will not lay at all. Most vaccinated subjects may be free from infection, but a number of unvaccinated may also escape. This sort of thing makes it difficult to know how much reliance can be placed on the averages; the results in a particular instance might be quite unpredictable. Thus, when comparing the clutch sizes in two species of birds, or the average attack-rates for inoculated and uninoculated subjects, we might be uncertain whether the differences observed were in some sense real or whether they were only due to chance variations. On the whole, one usually feels that conclusions based on large numbers of observations are more reliable than those based on small numbers. But the question still remains – how large a number is required for adequate reliability? Moreover, how does one measure this reliability?

To some extent, the expert worker in any field learns from experience how to deal with such difficulties. And the continued progress of biological science shows that he or she is not entirely unsuccessful in his or her efforts! However, it is frequently advantageous to try to use more precise methods of describing the basic variability, of deciding whether apparent differences are due to chance or not, of estimating unknown constants, and so on.

This is where one turns to statistical methods. Some people think that the great variation present in biological material makes statistical methods unreliable. In fact, very nearly the opposite is true. It is precisely because modern statistics is based on a recognition of this variation that it is such a powerful tool for handling numerical data. Great quantities of complicated experimental results can often be reduced to more manageable proportions by the calculation of a few numbers which characterise the whole pattern of events. Again, the application of probability theory, itself based largely on refinements of intuitive commonsense ideas, means that we can assess the odds that some apparent effect is or is not due to chance, or that the unknown true value of some constant lies between certain limits.

Now, it so happens that the application of statistical methods requires comparatively little mathematical knowledge or ability. The majority of procedures required in practice have been reduced to simple arithmetical calculations, most complications being avoided completely by the use of pocket calculators and computers.

The present book gives a range of elementary methods which should provide a biologist with what he or she is likely to require for perhaps 95 per cent of the time. Thus, a relatively small number of methods will do duty for a large number of situations. It is of paramount importance to understand the general conditions under which any particular method can be used. Statistical tests should never be applied automatically without first giving some thought to their validity. Again, it is more important to be able to recognise the occurrence of some non-standard situation than it is to be able to apply the proper method of analysis. Once the situation is recognised, expert statistical advice can be sought (preferably from a statistician with experience of biological applications, i.e. a biostatistician); if it is not recognised, erroneous and misleading methods may be used.

Another reason why it is useful for the biologist to have some familiarity with statistical ideas is the following. When mathematical difficulties arise, or when there is some doubt as to the best experimental design to be adopted, etc., the advice of a biostatistician will often be sought. Now, you may usually trust the professional biostatistician to work out correctly the theoretical consequences of the assumptions of a particular theory. The important point is to know whether the biostatistician is solving the right problem! If the biologist knows something of the general principles of statistical methods, he or she can cross-question the biostatistician who is helping him or her sufficiently closely to discover whether the latter is on the right track.

Although some of the methods used to deal with complicated experiments may be incomprehensible to the non-mathematician, the final results should be expressible in a form that the biologist can readily appreciate. It is usually safe to distrust any result that cannot, after a reasonable amount of discussion and explanation, be put in terms of the original biological problem. Sometimes it may appear that the wrong problem was tackled or the wrong questions asked, but in such cases adequate explanations ought to be forthcoming.

The general plan of this book is to show first, in Chapter 2, how statistics can describe and handle whole ranges of variation as opposed to mere averages. Then in Chapter 3 we see how more or less uncertain numbers, such as averages calculated from relatively limited samples of data, are related to the 'true' values we should get from extremely large samples. Chapter 4 introduces the basic idea of a statistical significance test, which is used to decide whether observed differences between factors are likely to be real or due only to chance. Chapters 5 and 6 then describe a number of the most frequently used significance tests. We next consider, in Chapters 7 and 8, ways of testing whether different groups of data are homogeneous, and whether experimental observations agree sufficiently closely with theoretical values for the theory to be regarded as reasonably adequate. After this, the problem of associated measurements is introduced, such as occurs when variation in one quantity, such as the length of an organism, is closely connected with variation in another quantity, such as the organism's weight. The simple case of a pair of measurements is dealt with in Chapter 9 on correlation and Chapter 10 on regression. More advanced methods of coping with several measurements at once are outlined in Chapter 14.

Chapters 11 and 12 discuss the important questions of how, in specific experiments, it is often possible to choose a pattern of experimentation that is not only highly informative but also leads to types of analysis that are simple to perform and interpret. In Chapter 13 we see how to avoid bias in collecting observations or in allocating experimental units. The important subject of nonparametric and distribution-free tests is introduced in Chapter 15, where a number of simple procedures are given for carrying out tests that do *not* depend on the distribution underlying the observations. The final chapter, Chapter 16, gives some advice on numerical work and the use of pocket calculators and computers with statistical software.

The succession of subjects introduced is intended to provide a graded course of instruction in elementary statistical methods.

Those who are already familiar with some of these may, of course, pick out any chapters in which they are specially interested. Most of the methods recommended are comparatively easy, though they may require a certain amount of practice before a real facility in their application is achieved. Only Chapter 14 is likely to present any real arithmetical difficulty. But the principles involved are important, and should be understood even if the labour of computation is avoided by using a calculator or computer.

An important feature of this book is the 'Summary of statistical formulae'. This is intended for use as a quick reference guide by the reader who already has some knowledge of statistics. It cannot be emphasised too strongly that standard formulae should not be applied blindly without some understanding of their suitability. Nevertheless, many workers who have already acquired a basic training in statistics, either from this book or elsewhere, will frequently require only to have their memories refreshed.

The seven appendix tables provided are to enable the reader to carry out the commonest statistical tests without special reference to more extensive compilations. There is, however, some advantage in possessing a good set of tables for general back-up. The best collection is probably Fisher & Yates' *Statistical Tables for Biological, Agricultural and Medical Research*. Another useful book is Volume I of *Biometrika Tables for Statisticians*, which contains a rather large number of tables, many of them not readily available elsewhere. An excellent and comprehensive collection of statistical tables may also be found in Ciba–Geigy's *Scientific Tables*. For day-to-day laboratory use the small and cheap *Cambridge Elementary Statistical Tables*, by D. V. Lindley & J. C. P. Miller, can be recommended. However, the need to use many of these tables is rapidly disappearing with the widespread availability of pocket calculators that will handle automatically all the common functions such as square roots, powers, reciprocals, logarithms, exponentials, trigonometric functions, etc. Moreover, many calculators and computer programs will also supply P-values for significance tests based on common distributions such as the normal, t, F, χ^2 etc.

Finally, a word should be said about further reading. This book attempts to provide the groundwork basic to most statistical methods. However, those workers who are closely concerned with special fields will want to know something about more advanced methods. It is possible in the subject of experimental design, for example, to learn to use relatively sophisticated patterns of experimentation without becoming involved in higher mathematics. To some extent the choice of text-books is a personal one, depending on the reader's own interests and way of looking at things. Specific recommendations are thus liable to be difficult. The section 'Suggestions for more advanced reading' therefore includes a variety of statistical books, some of them rather specialised, which the reader may find useful to consult.

2

Variability and frequency distributions

We have seen in Chapter 1 how considerable natural variation is inherent in the subject-matter of practically all biological and medical work. It cannot be effectively disposed of by taking a few averages and then regarding these as more-or-less precise measurements. We must learn to handle the whole pattern of variation as such. The present chapter introduces some of the more common patterns, and shows how these can be described in fairly simple numerical terms. A clear idea of the basic attitude involved in looking at one's data from this point of view is essential to a proper understanding of the elementary statistical methods recommended in later chapters.

2.1 THE NORMAL DISTRIBUTION

We shall begin by considering a simple continuously variable quantity such as stature. We know this varies greatly from one individual to another, and may also expect to find certain average differences between people drawn from different social classes or living in different geographical areas, etc. Let us suppose that a socio-medical survey of a particular community has provided us with a representative sample of 117 males whose heights are distributed as shown in the first and third columns of Table 1.

We shall assume that the original measurements were made as accurately as possible, but that they are given here only to the nearest 0.02 m (i.e. 2 cm). Thus the group labelled '1.66' contains all those men whose true measurements were between 1.65 and 1.67 m. One is liable to run into trouble if the exact methods of recording the measurements and grouping them are not specified exactly. In the example just given the mid-point of the interval labelled '1.66' is, in fact, 1.66 m. But suppose that the original readings were made only to the nearest 0.01 m (i.e. 1 cm) and then 'rounded up' to the nearest multiple of 0.02 m. We should then have '1.65', which covers the range 1.645 to 1.655, included with '1.66'. The interval '1.66' would then contain all measurements lying between 1.645 m and 1.665 m, for which the mid-point is 1.655 m. The difference of 5 mm from the supposed value of 1.66 m could lead to serious inaccuracy in certain types of investigation.

A convenient visual way of presenting data is shown in Fig. 1, in which the area of each rectangle is, on the scale used, equal to the observed proportion or percentage of individuals whose height falls in the corresponding group. The total area covered by all the rectangles therefore adds up to unity or 100 per cent. This diagram is called a *histogram*. It is easily constructed when, as here, all the groups are of the same width. It is also easily adapted to the case when the intervals are unequal, provided we remember that the *areas* of the rectangles must be proportional to the numbers of units concerned. If, for example, we wished to group together the entries for the three groups 1.80, 1.82 and 1.84 m, totalling 7 individuals or 6 per cent of the total, then we should need a rectangle whose base covered 3 working groups on the horizonal scale but whose height was only 2 units on the vertical scale shown in the diagram. In this way we can make allowance for unequal grouping intervals, but it is usually less troublesome if we can manage to keep them all the same width. In some books histograms are drawn so that the area of each rectangle is equal to the actual number (instead of the proportion) of individuals in the corresponding group. It is better,

Table 1. *Distribution of stature in 117 males*

Absolute height (m)	Working measurements with origin at 1.70 m (x)	Number of men observed (f)	Contributions to the sum (fx)	Contributions to the sum of squares (fx^2)
1.58	−6	1	−6	36
1.60	−5	3	−15	75
1.62	−4	6	−24	96
1.64	−3	8	−24	72
1.66	−2	13	−26	52
1.68	−1	18	−18	18
1.70	0	19	0	0
1.72	1	14	14	14
1.74	2	14	28	56
1.76	3	9	27	81
1.78	4	5	20	80
1.80	5	4	20	100
1.82	6	2	12	72
1.84	7	1	7	49
Total		117	+15	801

however, to use proportions, as different histograms can then be compared directly.

The general appearance of the rectangles in Fig. 1 is quite striking, especially the tall hump in the centre and the rapidly falling tails on each side. There are certain minor irregularities in the pattern, and these would, in general, be more pronounced if the size of the sample were smaller. Conversely, with larger samples we usually find that the set of rectangles presents a more regular appearance. This suggests that if we had a very large number of measurements, the ultimate shape of the picture for a suitably small width of rectangle would be something very like a smooth curve. Such a curve could be regarded as representing the true, theoretical or ideal distribution of heights in a very (or, better, infinitely) large population of individuals.

What sort of ideal curve can we expect? There are several theoretical reasons for expecting the so-called *Gaussian* or '*normal*' *curve* to turn up in practice; and it is an empirical fact

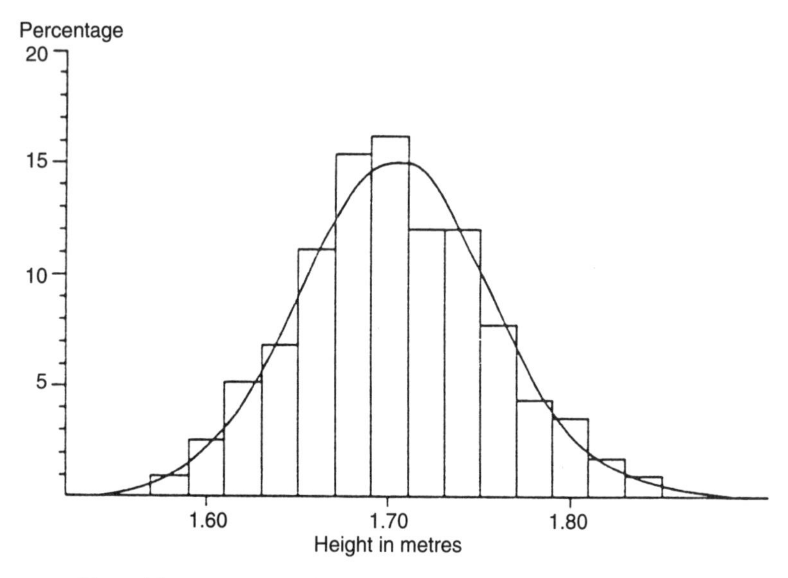

Fig. 1. The observed distribution of the heights of 117 males, exhibited in the form of a histogram (rectangles), together with a fitted 'normal' curve (smooth curve).

that such a curve often describes with sufficient accuracy the shape of histograms based on large numbers of observations. Moreover, the normal curve is one of the easiest to handle theoretically, and it leads to types of statistical analysis that can be carried out with a minimum amount of computation. Hence, the central importance of this distribution in statistical work.

The actual mathematical equation of the normal curve is

$$y = \frac{1}{\sigma\sqrt{(2\pi)}} \exp\left\{-\frac{(x-\mu)^2}{2\sigma^2}\right\}, \tag{1}$$

where μ is the *mean* or *average* value and σ is the *standard deviation*, which is a measure of the concentration of frequency about the mean. More will be said about μ and σ later. The ideal variable x may take any value from $-\infty$ to $+\infty$. However, some real measurements, such as stature, may be essentially positive.

But if small values are very rare, the ideal normal curve may be a sufficiently close approximation. Those readers who are anxious to avoid as much algebraic manipulation as possible can be reassured by the promise that no *direct* use will be made in this book of the equation shown. Most of the practical numerical calculations to which it leads are fairly simple.

Fig. 1 shows a normal curve, with its typical symmetrical bell shape, fitted by suitable methods to the data embodied in the rectangles. This is not to say that the fitted curve is actually the true, ideal one to which the histogram approximates; it is merely the best approximation we can find.

The normal curve used above is the curve we have chosen to represent the *frequency distribution* of stature for the ideal or infinitely large *population*. This ideal population should be contrasted with the limited *sample* of observed values that turns up on any occasion when we make actual measurements in the real world. In the survey mentioned above we had a sample of 117 men. If the community were sufficiently large for us to collect several samples of this size, we should find that few if any of the corresponding histograms were exactly the same, although they might all be taken as illustrating the underlying frequency distribution. The differences between such histograms constitute what we call *sampling variation*, and this becomes more prominent as the size of sample decreases.

It might happen, however, that there were only 117 men in the whole community. Although it would not then be possible *in fact* to draw and compare several different samples, we should still find it convenient to regard the 117 men observed as a single sample from some ideal or hypothetical population.

We shall see later how certain important characteristics of the population can be inferred from calculations performed on samples. For the moment, however, we are confining our attention to clarifying the general idea of frequency distributions.

Definition of probability

This is perhaps a suitable point at which to say something about the meaning of 'probability'. It is unnecessary in a book of this kind to make heavy weather with logical subtleties, although of course the concept is both basic and important.

Probability is best defined here in relation to the frequency distribution of the ideal population. Suppose we knew, for example, the true normal curve describing variations in height. Then the probability that an individual chosen at random had a stature between, say, 1.62 and 1.64 m would be precisely the area lying under the curve and above the x axis between two vertical lines drawn through the points $x = 1.62$ and $x = 1.64$, measuring x here in the units actually observed. This area represents the relative frequency or probability with which we expect such statures to turn up 'in the long run', that is, in very large samples, for which the observed histogram should be close to the theoretical curve.

The above definition fits in quite well with intuitive common-sense ideas of probability. At the same time, it should be noticed that the definition relates to a state of affairs in which we envisage, at least in principle, the indefinite repetition of some measurement. Many statisticians prefer to restrict the use of 'probability' to such situations, and would not speak of 'the probability that a scientific hypothesis or theory is true'. According to the restricted usage, hypotheses and theories, however credible, 'probable' or 'well-established', are not experimentally repeatable observations from some ideal population of values. There is considerable controversy on this point, especially in regard to the use of the currently popular 'Bayesian' approach, though elementary statistical practice is largely unaffected.

2.2 THE BINOMIAL DISTRIBUTION

Having considered the simplest example of variables that are most conveniently thought of as basically continuous, we now turn to the kind of observations that are essentially discontinuous or discrete. An example will make the distinction clear.

Suppose that in a human pedigree involving albinism, which is a simple recessive, we find a marriage in which both partners are known to be heterozygous for the albino gene. This is an 'intercross' for which, on Mendelian theory, we expect one-quarter of the children to be albinotic. We say that the chance that any child will be affected is one-quarter, or that 'on average' or 'in the long run' the proportion of albinotic children is expected to be one-quarter. Any given 'run' is, of course, in practice never very long. Moreover, a family may consist of any possible combination of normals and albinos, though some patterns are more likely than others.

Again we have a considerable amount of variation in any data that we are likely to come across. Nevertheless, we can expect this variation to have a certain well-defined pattern, which will be described by an appropriately chosen distribution. Let us see what will happen with families having just two children. These may be both normal, one normal and one affected, or both affected. Since the chance that any particular child is normal is $\frac{3}{4}$, the chance that both are normal is $(\frac{3}{4})^2 = \frac{9}{16}$, using the rule that the joint probability of two *independent* events is the product of their individual probabilities. Similarly, the chance of both being albinos is $(\frac{1}{4})^2 = \frac{1}{16}$. The balance of probability, namely $1 - \frac{1}{16} - \frac{9}{16} = \frac{3}{8}$, corresponds to one normal and one affected. A general formula for calculating these probabilities is given below in formula (2). These three probabilities constitute the whole of the theoretical frequency distribution, for which the variable number of albinos is not continuous but can take only the specific values 0, 1 or 2.

If we managed to collect a sample of 12 such families, we might find that 6 contained all normals, 5 one normal and one affected, and 1 two affected. The expected or theoretical long-run values would be $\frac{9}{16} \times 12$, $\frac{3}{8} \times 12$, and $\frac{1}{16} \times 12$, or $6\frac{3}{4}$, $4\frac{1}{2}$ and $\frac{3}{4}$. The comparisons are shown in Table 2. The 12 families observed can be thought of as a small sample from a very large population of families whose theoretical frequencies are those given by the calculations described above. The population exhibits, in fact, a *binomial distribution* of probabilities, the name being derived from the fact that the individual frequencies are given by the terms of the binomial expansion of $(\frac{3}{4} + \frac{1}{4})^2$.

The binomial distribution arises whenever there are two alternative characters which an individual may have – here normality and albinism. Extensions can be made to several characters, and

Table 2. *Distribution of albinotic children in 12 families of 2*

	Number of albinotic children			
	0	1	2	Total
Observed number of families	6	5	1	12
Expected number of families	6.75	4.50	0.75	12

this leads to the use of *multinomial* distributions. Text-books usually give examples drawn from games of chance involving dice or cards. Biological instances are very common. There are many genetical applications, not only of the type described above, but also to the more complex problems of detecting linkage. Again, we frequently need to compare percentages, such as the attack-rates for inoculated and uninoculated groups of subjects. In any analogous situation, the basic frequency distribution can usually be expected to be binomial. It is more important to recognise this state of affairs than it is to know much about the theory of the distribution. As we shall see later, most statistical applications can be carried out by using simple routine formulae.

In the illustration above, we have considered only a very special case of a binomial distribution. More generally, we suppose that the individuals examined may possess a certain character with probability p, or may fail to exhibit it with probability $1 - p = q$. The probabilities of a sample of n individuals having 0, 1, 2, . . . , n individuals possessing the character in question are given by the terms of the binomial expansion of $(q + p)^n$. This means that the chance that exactly a individuals bear the character is

$$\frac{n!}{a!(n-a)!} p^a q^{n-a} \tag{2}$$

for $a = 0, 1, 2, \ldots, n$. We rarely need to use this formula as such in elementary applications, but it is necessary if we want to find the component probabilities.

In the illustration above we were thinking of families of two children, i.e. $n = 2$, where the chance of any child being an albino was one-quarter, i.e. $p = \frac{1}{4}$.

2.3 THE POISSON DISTRIBUTION

Besides the normal and binomial distributions just described, there is a third which frequently arises in practice in certain circumstances. A typical situation is that which occurs when a haemocytometer is used for counting cells with a view to estimating the density of the suspension from which they are taken. The main feature to be emphasised here is that although there is an exceedingly small chance of any particular cell finding its way to a given small square of the haemocytometer, the total number of cells is so large that quite a lot of the squares are likely to be occupied by one or more cells. When these conditions apply, we expect the number of cells in a square to have a *Poisson distribution*. The figures shown in Table 3 relate to data originally collected by 'Student' on the counting of yeast cells. There were 400 small squares in all, and the table shows how many of these were occupied by 0, 1, 2, . . . cells.

Table 3 also gives the average number of squares expected in each class from a fitted theoretical Poisson distribution. There is, as we should expect, some variation between the observed numbers and the theoretical ones, but the general impression is that there is reasonably good agreement.

A very similar state of affairs arises when we are dealing with the growth of bacterial colonies on a plate of culture medium. A famous and often quoted historical example is that of Bortkiewicz, who discovered that the number of men kicked to death by a horse in a Prussian army corps each year closely followed a Poisson distribution. Another well-known Poisson variable is the number of radioactive atoms that disintegrate in a given volume during a given time.

The mathematical description of the Poisson distribution is as follows. The chance of obtaining exactly x individuals in typical conditions, e.g. cells in a haemocytometer square, is

$$\frac{e^{-m} m^x}{x!} \qquad (3)$$

for $x = 0, 1, 2, \ldots$, and so on without limit. It will be noticed

Table 3. *Distribution of yeast cells in a haemocytometer (Student's data)*

Number of cells in a square (x)	Observed number of squares	Expected number of squares
0	0	3.7
1	20	17.4
2	43	40.6
3	53	63.4
4	86	74.2
5	70	69.4
6	54	54.2
7	37	36.2
8	18	21.2
9	10	11.0
10	5	5.2
11	2	2.2
12	2	0.9
Greater than 12	0	0.4
Total	400	400.0

that this distribution involves only one parameter, m, which alone determines the shape of the whole distribution.

2.4 OTHER DISTRIBUTIONS

We have introduced above the three commonest and most useful frequency distributions. Others of quite different character may of course easily arise in practice. Thus, the distribution of length of life is often a reversed J-shaped curve starting with its highest point at or near the origin and steadily dropping with increasing age. Again, the distribution of cloudiness, measured on the usual points scale, is frequently U-shaped, with concentrations of frequency near the points for 'clear sky' and 'completely overcast'. When such distributions occur, it is best to seek expert statistical advice. The main point is to be able to recognise when the simpler distributions dealt with in this book do not apply and cannot be regarded as sufficiently good approximations to the situation under investigation.

2.5 MEANS AND VARIANCES: BASIC CALCULATIONS

In the previous sections we have seen how various frequency distributions can be used to describe different types of variation commonly found in biological and medical data. Although the mathematical forms of the distributions were given, it was pointed out that it was usually unnecessary in practice to know much about the underlying theory. There are, fortunately, a few simple methods of describing and handling such distributions based on numbers that can readily be calculated from the data observed.

The normal distribution

It is convenient to take the normal distribution first. This distribution is determined by just two quantities, μ and σ. The mean μ is simply the average value of the ideal population, and is the value of x corresponding to the highest point of the smooth symmetrical curve shown in Fig. 1.

The *standard deviation* σ is a measure of the dispersion of probability about the highest point. If σ is small, the distribution is tightly packed round the central value and large departures occur only very rarely. Alternatively, if σ is large, the distribution is rather flat and widely spread: great natural variation is present and considerable departures from the mean are quite common. In fact, using an electrical analogy, we can say that σ is the 'root-mean-square about the mean', i.e. the square root of the average value of $(x - \mu)^2$ taken over the whole of the ideal population. For the kind of normal curve shown in Fig. 1, σ is the distance on the x-axis between the values corresponding to the peak of the curve and to the points (one on each side) of greatest slope. Quite often we find it more convenient to work with the *variance* σ^2, which is simply the square of the standard deviation. This is, therefore, the 'mean-square about the mean'.

Now, it may well happen that we do not know the actual values of μ and σ because they relate to the ideal population, of which we can observe only rather variable samples of instances. Nevertheless, as we shall see later in dealing with significance tests, scientific hypotheses may assume specific values of such unknown constants, e.g. that the true difference in the average heights of the males and females of a certain species is in fact zero.

In general, therefore, we can obtain only more-or-less inaccurate *estimates* of μ and σ from the sample of data available to us. If we make these estimates in the right way, we usually find that we obtain better and better approximations to the true μ and σ as the samples become larger. The important question is then – 'how do we calculate suitable estimates in any particular case?'

The required rules are specified most simply by two or three algebraic formulae, which will first be explained in general terms, and then illustrated by application to the data in Table 1. We shall suppose that we have a sample of n observations, each of which may be represented generally by the symbol x. (It is sometimes convenient to distinguish the individual observations by means of a suffix, e.g. $x_1, x_2, \ldots, x_n$, but this complication is unnecessary in the present context.) The mean or average of the n observations, which we shall call $\bar{x}$, is thus given by

$$\bar{x} = \frac{1}{n}\sum x. \tag{4}$$

The mathematical summation symbol $\sum$ simply means that all instances of whatever directly follows it must be added together, and here we are restricting ourselves to the instances of x provided by the sample. (We could, therefore, have written, more elaborately,

$$\bar{x} = \frac{1}{n}(x_1 + x_2 + \ldots + x_n) = \frac{1}{n}\sum_{i=1}^{n} x_i.)$$

The important thing to remember is that although any observed $\bar{x}$ is unlikely to be exactly equal to the unknown μ, and indeed may depart from it appreciably, it is the best and most convenient *estimate* of μ available to us.

We now turn to an estimate of the unknown σ. As mentioned above, σ is the 'root-mean-square about the mean' in the population. A suitable estimate s can be derived from a sample by taking something very close to the root-mean-square about the *sample* mean $\bar{x}$. Thus, for the estimate of variance we use

$$s^2 = \frac{1}{n-1}\sum(x - \bar{x})^2, \tag{5}$$

where now we are summing the squares of the deviations of all observations from their mean; the standard deviation s is of course simply the square root of the right-hand side of formula (5). It is important to notice that the divisor in formula (5) is $n - 1$ and not n; the difference can be substantial if n is small. If in a particular case we knew the population mean μ but not σ, we should estimate the latter from the square root of

$$s^2 = \frac{1}{n}\sum(x - \mu)^2, \tag{6}$$

where the divisor is now n instead of $n - 1$.

Except when samples are fairly small, it is best to calculate $\sum(x - \bar{x})^2$ not in the direct and obvious way of first finding $\bar{x}$, subtracting this from each x, squaring all the results thus obtained and adding – a laborious procedure – but by employing the identity

$$\sum(x - \bar{x})^2 = \sum x^2 - \frac{1}{n}\left(\sum x\right)^2. \tag{7}$$

The advantage of this is that we work immediately with the observed values and their squares. Approximation comes only when $(\sum x)^2$ has to be divided by n. Whereas if we use $\sum(x - \bar{x})^2$, not only does every difference $x - \bar{x}$ have to be calculated separately, but each of these differences also contains an error due to the approximation in $\bar{x}$.

The actual calculation of $\bar{x}$ and s is readily illustrated by the data in Table 1. The arithmetic has been reduced somewhat by choosing a *working origin* at 1.70 m, so that x is used for deviations from this value. Another point to be noticed is that with grouped data each value of x may appear several times, say with a frequency f. The basic quantities required could thus be written $\sum fx$ and $\sum fx^2$, although it is unnecessary to include f if we understand that every value x is to be counted as many times as it appears in the sample.

The table should be self-explanatory. The second column is in working measurements derived from the first column by adopting 1.70 m as a working origin, and using units of 0.02 m. The third column gives the observed numbers of men in each height group. Calculations must be performed to fill in the rest of the table. Each entry in column 4 is the product of the corresponding numbers on the same line in columns 2 and 3. And the entries in column 5 are formed from the products of the numbers in columns 2 and 4. Finally, we total each of columns 3, 4 and 5. Arithmetic must be carefully checked. Care must also be taken with the signs in column 4, but all entries in column 5 must be positive.

We now have the main ingredients required for the final calculations, namely $n = 117$, $\sum x = +15$ and $\sum x^2 = 801$. The mean, measured in working units from the working value of 1.70 m, is

$$\bar{x} = \frac{1}{n}\sum x = \frac{15}{117} = 0.128$$

so that the mean on the ordinary scale of measurement is $1.70 + (0.128 \times 0.02) = 1.703$ m. For the variance we need first the sum of squares about the mean. Formula (7) gives

$$\sum(x - \bar{x})^2 = 801 - \frac{15^2}{117} = 799.08.$$

We next apply formula (5), dividing by $n - 1 = 116$, to obtain s^2. Thus in working units we have

$$s^2 = \frac{799.08}{116} = 6.889,$$

with

$$s = \sqrt{6.889} = 2.625.$$

In the ordinary scale of measurements, we have

$$s = 2.625 \times 0.02 \text{ m} = 0.0525 \text{ m}.$$

In practice, of course, using a pocket calculator with a built-in statistics facility, we need only feed in the string of x-values: pressing the relevant keys then yields $\bar{x}$ and s immediately.

The above simple calculations are all that are, in general, required to obtain the sample estimates $\bar{x}$ and s of the ideal population values μ and σ. Although the basic formulae really apply to the measurements x as originally made, we have in Table 1 condensed this basic material by use of groups one working unit, or 0.02 m, in width. The mean value is on average unaffected by this procedure, but the variance tends to come out a little too high. It is customary therefore to use *Sheppard's correction to the variance*, and subtract one-twelfth of the square of the group interval from the value calculated as shown above. The corrected variance in working units squared is then $6.889 - 0.083 = 6.806$, giving an adjusted standard deviation of working units, or $2.609 \times 0.02 \text{ m} = 0.0522 \text{ m}$ in the units of actual measurement. The general formula for the corrected variance s'^2 is

$$s'^2 = s^2 - \tfrac{1}{12}h^2, \tag{8}$$

where h is the group interval employed.

The smooth curve in Fig. 1 is obtained by substituting in formula (1) the estimates of μ and σ just calculated, and then multiplying by 117 to make the totals agree. Such curves have various characteristic properties, and although detailed mathematical knowledge is unnecessary it is useful to bear in mind one or two salient features. The symmetrical bell shape of the distribution is striking. Since the probability is concentrated round the mean, departures from this value become progressively rarer as we move away from it. In deducing the practical implications of such normal distributions it is worth remembering that about 32 per cent of the total probability lies outside a range extending one standard deviation each side of the mean. Further, a range of 1.96 standard deviations in each direction excludes only 5 per cent of the distribution, i.e. $2\frac{1}{2}$ per cent in each 'tail'. Thus, in the numerical example above we should expect approximately 95 per cent of individuals to have heights in the range $1.703 \pm (1.96 \times 0.0522)$, i.e. 1.601 to 1.805 m. This type of calculation depends on the sample size being reasonably large, say $n > 30$. If n is smaller than this, we must resort to the methods of section 6.4. A number of multiplying factors like 1.96, corresponding to different values of the excluded probability, are given in Appendix 1.

We have in the foregoing discussion been thinking solely in terms of calculating the mean and standard deviation of a *normal* curve which we suppose might adequately describe the data. Although the calculations indicated are of greatest value in connection with normal distributions, the definitions of mean and standard deviation apply to any distribution. However, unless alternative curves are well understood, there may be some difficulty in interpreting quantities like the standard deviation. For example, it will in general no longer be true that a range of 1.96 standard deviations excludes 5 per cent of the total probability.

The binomial distribution

The binomial distribution is in many ways simpler. We saw in section 2.2 how the frequencies may be calculated for a sample of n observations, when the chance that an individual bears a particular character is p. In the genetical example used, we were examining the possibility that $p = \frac{1}{4}$. More generally, p will be unknown and must be estimated from the data. Thus Table 2 involves 7 affected children amongst the 24 in the 12 families. An estimate of the segregation ratio is therefore $\frac{7}{24} = 0.29$. Again, if we are making a differential white cell blood count and find that 63 cells out of 100 are polymorphonuclear, then the estimated chance that a given cell is of the latter type is simply $\frac{63}{100} = 0.63$. The full distribution for this value of p in samples of 100 would then be given by formula (2) with $p = 0.63$ and $n = 100$, with a varying from 0 to 100.

If we consider the distribution of a represented by formula (2), it can be shown that the average is given by $\mu = np$, as we should expect, and that the variance is $\sigma^2 = npq$. Further, the estimate

a/n has mean p and variance pq/n. These expressions are important and worth remembering.

In particular, if n is fairly large and p not too small, the distribution of a tends to a normal distribution with mean np and variance npq. We have a similar result for the estimate a/n, which tends to have a normal distribution with mean p and variance pq/n. This simplifies the interpretation of results, as it means we can avoid calculation of the individual probabilities given in formula (2). More will be said about this in the following chapter on estimation.

The Poisson distribution

The Poisson distribution is also dealt with quite easily, because we simply use the sample average of the variable involved to estimate the Poisson parameter m appearing in formula (3). Consider the data of Table 3. There are 400 squares in all, the number of cells in each square being the variable in question. The total number of cells is 1872 with an average of $\bar{x} = 4.68$ per square. We can therefore estimate m as 4.68. Substituting this quantity in formula (3) gives the estimated probabilities with which squares contain different numbers of cells. (Alternatively, we can make use of a table in *Biometrika Tables for Statisticians*, which gives individuals terms of the Poisson series for different values of m.) Multiplying each probability by 400 yields the theoretical numbers of squares expected. These are shown in the third column of Table 3. Comparison of columns two and three reveals quite good agreement between the observed and the theoretical values.

The variance of a Poisson distribution is also the quantity m, so that the standard deviation is $\sqrt{m}$. The Poisson tends to normality as m increases, so that once again we can make use of the normal curve in drawing general conclusions. If m is small, however, the curve tends to be rather skew with a maximum near the origin and a long tail.

3

Estimation, standard errors and confidence limits

3.1 SAMPLING VARIATION

We have in Chapter 2 made the acquaintance of three important types of statistical frequency distribution – the normal, the binomial and the Poisson. As explained, any set of data will usually be, at best, only a sample of some theoretical population. Thus, in Fig. 1 we have the histogram formed by the heights of a sample of 117 males. The smooth curve in the diagram is the best-fitting normal curve, which may be regarded as describing an ideal underlying distribution of stature. The ideal curve is unknown, in this example at any rate, but with a reasonable number of observations we may expect to obtain quite a good approximation to it. This means we must be content to use the sample estimates $\bar{x}$ and s in place of the unknowns μ and σ. With the binomial distribution we have to estimate the unknown probability of 'success' p by the proportion observed in a sample of size n. Similarly the unknown Poisson m must be estimated by the sample mean $\bar{x}$.

Now it will easily be realised that even if the underlying distribution remains the same, the estimates of parameters provided by repeated independent samples will vary. It just happens that the particular set of observations available to us gives a certain value of $\bar{x}$; another set of observations might yield quite a different value of $\bar{x}$, the difference being due entirely to chance variation. Fortunately, such discrepancies usually become smaller as more observations are available, and we then say that the estimates are more reliable or precise.

Suppose that it were possible, as indeed it sometimes is in practice, to collect repeated independent samples of data and to calculate from each of these such quantities as $\bar{x}$ or s. Then it is clear that we can legitimately speak of the frequency distribution of these estimates themselves. Such distributions are, in general, different from those of the individual observations. These sample distributions are very important as they afford measures of precision for the estimates concerned. Sometimes the sample distribution of an estimate is approximately normal. The standard deviation of this distribution is then a direct measure of precision, and the smaller this standard deviation the more reliable the estimate.

Before considering some specific distributions, we shall first take note of two general results for the behaviour of sample means. The first is that if we consider the distribution of $\bar{x}$, supposed to be calculated for repeated samples each of n observations, then the mean of this distribution is also μ, i.e. the same as the mean of the individual observations. This is fortunate and also expected. It means that whenever we use $\bar{x}$ for estimating μ, the various values of $\bar{x}$ would in the long run be centred on μ and would not tend to cluster to one side or the other.

The second result required deals with the variance of the distribution of $\bar{x}$. Whereas the original distribution of each observation has variance σ^2, the variance of $\bar{x}$ calculated from a sample of observations is σ^2/n. The denominator n leaps to the eye; it means that as n increases, the variance of $\bar{x}$ decreases in direct proportion to n. The corresponding standard deviation is of course $\sigma/\sqrt{n}$. It follows that in the sampling distribution of $\bar{x}$ the different values that might occur are much more closely packed about the true value μ than are the single observations of the basic distribution. This is the chief reason why the mean values of samples give a more reliable indication of the true population mean than do single isolated measurements. We shall now consider in more detail how these ideas apply to specific distributions.

3.2 THE NORMAL DISTRIBUTION (LARGE SAMPLES OF SIZE $n > 30$)

We first look at the normal distribution. In Chapter 2 data were given on the stature of 117 males. We found the actual sample mean (measured in metres, rather than in deviations from a working mean of 1.70 m) to be $\bar{x} = 1.703$ m, and the corrected standard deviation to be $s = 0.052$ m. It was pointed out that if these values of $\bar{x}$ and s were taken as estimates of μ and σ, then we should expect 95 per cent of actual heights to lie in the range 1.601 to 1.805 m, an interval of just over 0.2 m.

Now let us use the result, given in the previous section, that the variance of $\bar{x}$ in repeated samples is σ^2/n, with corresponding standard deviation $\sigma/\sqrt{n}$. We estimate this by $s/\sqrt{n} = 0.052/\sqrt{117} = 0.005$, an appreciably smaller quantity than the original s. An important and useful property of the normal distribution is that mean values are also normally distributed, with mean μ and standard deviation $\sigma/\sqrt{n}$. The appropriate distribution of $\bar{x}$ in the present case is therefore approximately normal, having estimated mean 1.703 and estimated standard deviation 0.005. In large samples, with $n > 30$, we can calculate 95 per cent limits for the mean on the basis of a range of 1.96 estimated standard deviations each side of the mean, i.e. $1.703 \pm (1.96 \times 0.005) = 1.693$ to 1.713 m, an interval of only 0.02 m (i.e. 2 cm).

These results can be presented in various ways. One method is to quote the estimated mean with its standard deviation, or *standard error* as it is often called in this context, e.g. 1.703 ± 0.005 m. Symbolically, we write

$$\bar{x} \pm s/\sqrt{n}. \tag{9}$$

When doing this, it is advisable to indicate that one is giving the standard error of the mean, in case the figure following the '±' sign is mistaken for the standard deviation of the individual measurements. (In some of the older literature it was customary to provide a *probable error*. The range thus entailed covers 50 per cent of the total probability. The probable error is 0.6745 times the standard deviation, and its use is best avoided.)

Alternatively, one can work directly in terms of suitable limits, e.g. the 95 per cent range for the unknown mean calculated above as 1.693 to 1.713 m. Such limits are called *confidence limits*. A 95 per cent confidence interval strictly means that, if we assert that the true unknown mean μ lies in this interval, then we shall be correct (on the assumptions made) for 95 per cent of the time. Since μ is fixed though unknown, it is the interval and not μ which has a probability distribution for repeated sampling. Speaking rather loosely, however, it is sufficient to interpret the 95 per cent limits as meaning that the odds are '19 to 1 on' that the true value lies in the interval.

The use of confidence limits is more general than that of standard errors. The latter are readily interpreted, usually in terms of suitable confidence limits, only when the quantity to which they are attached is approximately normal in distribution. Confidence limits can often be found for a much wider class of distributions; we shall come across examples later in dealing with the t-distribution.

Although we have developed the above discussion in terms of 95 per cent limits, similar arguments are available for any other range that we prefer to work with, say 90 per cent or 99 per cent. Some appropriate multipliers for the standard error in the case of normal distribution can be read from Appendix 1.

General expressions for confidence limits for the mean of a normal distribution based on a reasonably large sample (say $n > 30$) are

$$\bar{x} - ds/\sqrt{n} \quad \text{to} \quad \bar{x} + ds/\sqrt{n}, \tag{10}$$

where d is chosen from Appendix 1 to correspond with the required probability. Thus a 95 per cent range excludes 5 per cent of probability, for which $d = 1.96$.

A useful result, to which we shall have cause to appeal again later, is that mean values often tend to be normally distributed, especially when based on many observations, even when the distribution of the basic measurements is not normal. This implies that standard errors can be meaningfully attached to such mean values, and interpreted accordingly.

3.3 THE BINOMIAL DISTRIBUTION (LARGE SAMPLES OF SIZE $n > 30$)

We shall now consider the application of the general ideas in the previous section to the binomial distribution. The way in which this distribution arises has been described in Chapter 2. Typically, we examine n observational units of which a are found to have some specific property, e.g. are polymorphonuclear cells in a differential white cell count. We suppose that the chance of any cell being polymorphonuclear is $p(=1-q)$, and use the ratio a/n to estimate p.

Now, as already stated, the distribution of a/n in repeated samples has mean p and variance pq/n. Moreover, in reasonably large samples the distribution is approximately normal. We can, therefore, apply the results of the previous section, presenting the estimate together with a standard error in the form

$$\frac{a}{n} \pm \sqrt{\left[\frac{\frac{a}{n}\left(1-\frac{a}{n}\right)}{n}\right]}. \qquad (11)$$

If, to use a previous illustration, we counted 63 polymorphonuclear cells in a total of 100 white cells, we should have $a = 63$, $n = 100$. Formula (11) then gives the resultant estimate of p as 0.630 ± 0.048. The usual 95 per cent limits would be 0.536 to 0.724. In practice we might as well consider this range too wide for reliable diagnostic use. Clearly a larger sample would then be required. Since the standard error involves the factor $1/\sqrt{n}$, we should have to increase the sample size by a factor of 4 in order to halve the standard error and the corresponding confidence interval. A similar calculation could be made for any other reduction we might have in mind.

When the binomial samples are small, say with n less than 30, the validity of using a normal approximation is more doubtful, especially if p is at all near 0 or 1. Approximate confidence intervals can still be calculated though this is mathematically more difficult. Fortunately some convenient charts are available from which the requisite limits can be read directly (see *Biometrika Tables for Statisticians*).

3.4 THE POISSON DISTRIBUTION

The Poisson distribution can be treated in a manner similar to that accorded the binomial in the last section. We have already seen in Chapter 2 how the Poisson parameter m can be directly estimated by the mean $\bar{x}$ of a sample in which we have n measurements, each expected to follow a Poisson distribution. Moreover, the standard deviation of the ideal population is $\sqrt{m}$. When we come to the problem of finding confidence intervals for m, our path is again smoothed by the fact that when m is large the Poisson distribution tends to normality. And even when m is not large $\bar{x}$ may be approximately normal provided the sample is large. Provided the product $n\bar{x}$ is larger than about 30, we can estimate m and attach a standard error by means of the expression

$$\bar{x} \pm \sqrt{\frac{\bar{x}}{n}}. \tag{12}$$

In the example where yeast cells were being counted in a haemocytometer there were $n = 400$ basic measurements, for which the average was 4.68. The average number of cells per square could therefore be estimated as

$$4.68 \pm \sqrt{\frac{4.68}{400}} = 4.68 \pm 0.11,$$

so that m would be known with some accuracy, since the 95 per cent range would be 4.46 to 4.90.

If, on the other hand, we have a single plate of culture medium with 30 bacterial colonies growing on it, this is only a single observation in terms of the previous discussion. Thus $n = 1$, and our estimate of the ideal value relevant to the conditions of the experiment is $30 \pm \sqrt{30} = 30 \pm 5.48$. The 95 per cent range is 19.3 to 40.7. Although $n = 1$ is small, $n\bar{x} = 30$ is just about large enough for us to make use of the normal approximation to the Poisson. Tables are available giving confidence limits for m for small values of Poisson variables (see *Biometrika Tables for Statisticians*). With a single reading of 30, the tables give 95 per cent limits of 20.2 to 42.8, which are not very different from the approximation above. Single values like this are, however, rarely very precise, but may sometimes yield sufficient information for practical purposes.

4

The basic idea of a significance test

Ada, Agnes, Alice, Amy, Angela . . . Each year or two Mrs Smith has a new baby, and so far all five children have been girls. Not a very surprising occurrence perhaps, for if you look far enough afield you will find a few families with even larger numbers of children all of the same sex. Still, it will undoubtedly excite a certain amount of local comment amongst neighbours and relatives who are more interested in Mrs Smith and her children than in the composition of other more-or-less remote and unknown families. A biologist might well want to assess more exactly the view that Mrs Smith is prone to produce only girls, perhaps with an eye on some possible explanation such as the inviability of Mr Smith's Y-bearing spermatozoa. Moreover, if Mrs Smith is anxious to have a son, the family doctor may be faced with some difficult problems in medicine, ethics and arithmetic. At the present stage a cautious person might be unwilling to say more than: 'We can't really be sure yet that there is anything unusual happening; there's probably a fifty-fifty chance that the next child will be a boy'. But suppose the series continues with Arabella and Augusta – the evidence for some abnormality is clearly stronger. To take a logically similar situation, there comes a point when you suspect that the person who repeatedly spins his or her own coin and correctly calls heads each time is either cheating in his or her calling or has a heavily biased or even two-headed coin.

Now, if we want to deal with such problems in a reasonably rational and objective manner, it will be necessary to adopt some standard method of making a judgement. Otherwise we shall be unduly swayed by prejudice: the enthusiastic seeker after abnormality will be impressed by, say, only three girls in a row; while the sceptic can always fall back on the comment that the one-in-a-million chance is bound to turn up somtimes, however many successive births are all of the same sex. If the first approach is adopted in scientific work, it results in too much effort being wasted in following up hunches that lead nowhere, while the second attitude stifles research by an excessive unwillingness to abandon existing theories.

A numerical technique which has been widely used for many years with great success is the *statistical significance test*. This takes various forms according to the nature of the problem with which we are dealing. Nevertheless, the underlying logical procedure is always the same. Although philosophical defences can be constructed, the best justification for such methods, at least so far as the research worker is concerned, lies in their proved practical utility. The type of reasoning employed is as follows.

We first consider, in relation to any particular situation or experiment, what statisticians call the *null hypothesis*. This is merely a general description of what we expect to happen according to some standard scientific theory, either before any observations have occurred or, at any rate, before existing data are looked at in detail. What we have in mind is that we shall stick to the standard theory unless the observations are so unexpected as to force us to abandon or modify it. This is entirely in accord with the ordinary scientific practice. In the above illustration of Mrs Smith's children a convenient null hypothesis would be that there was a fifty-fifty chance (ignoring here the known small departure of the sex-ratio at birth from this value) of each child being a girl (or boy), and that the sex of any child was uninfluenced by the sex of any previous child.

The next step in carrying out a significance test is to calculate the probability of the actual event observed, *together with any other equally extreme or more extreme events that might have occurred*. The last italicised phrase is important. It often happens that each possible result of an experiment is by itself fairly improbable. So the probability of what we observe, taken by itself, may prove nothing. What is of greater relevance is a suitable group of results, all of which show extreme deviations from what we might as a rule expect. Further discussion of the illustration will make this clear.

Suppose we consider a family of just five children, and assume the null hypothesis suggested above. Then the number of girls involved will follow a binomial frequency distribution, as already described in Chapters 2 and 3. (And in the notation of those chapters we shall have $n = 5$ and $p = \frac{1}{2}$.) The most likely numbers to turn up in practice are two and three, while five or none are the most extreme values in the sense of being farthest removed from the 'centre' of the distribution. Since the chance of each child being a girl is one-half, the chance of all five being girls is $(\frac{1}{2})^5 = \frac{1}{32}$, or about 3.1 per cent. Now, as stated above, we must also take into consideration any other possibilities which are equally or more extreme. In this example there is only the event of *no girls*, which also contributes 3.1 per cent to the probability required. The latter is therefore 6.2 per cent, and is often called 'P' or the 'P value'.

We have now reached the core of a significance test. If the probability P is fairly large, say 30 per cent or more, then clearly the event which has actually occurred is one of a type that is not at all uncommon. And there is accordingly no ground for supposing that there is much wrong with the null hypothesis, which we can continue to accept. If, on the other hand, P is very small, say $\frac{1}{1000} = 0.1$ per cent or less, then we are faced with a crucial choice: *either* the null hypothesis is true as assumed and a very improbable type of event has occurred, *or* the null hypothesis is false. When the type of event that occurs is sufficiently improbable if the null hypothesis is true, we go for the second alternative and reject the null hypothesis. This has the effect of making the event in question more likely.

The secret of using this procedure successfully is to know

where to draw the line between what is and what is not sufficiently improbable for the null hypothesis to be rejected. There is a good deal of latitude permissible here, and the choice of where to draw the line, or the *significance level* used, depends to some extent on circumstances. Nevertheless, it has been found convenient in general scientific work to adopt the 5 per cent level for ordinary tests of significance. Now, in the example we found P to be 6.2 per cent. Since this is larger than 5 per cent, we should judge the data to be non-significant, and should say that there was insufficient evidence for rejecting the null hypothesis at this level.

If, on the other hand, there had been six children who were all girls, the corresponding calculation would have led to $P = 3.1$ per cent. This arises from the fact that the chance of exactly six girls is $(\frac{1}{2})^6 = \frac{1}{64}$, and we must also include the other equally extreme possibility of six boys. Since this value of P is less than 5 per cent, we should say that the result was significant at the 5 per cent level. The null hypothesis would therefore be rejected. If we still kept to the view that the sex of any child was unaffected by the sex of any previous children, the consequence of obtaining a significant result would be that we could no longer assume that the chance of any child being a girl was one-half. This would be equivalent to accepting the view that something had happened that required special explanation. Since the probability value of one-half would then be unacceptable, we might go on to ask what value or values would be compatible with the data. We could hardly expect any very precise answer to this question on the basis of only one family of six children. Nevertheless, something could be said using the method of confidence limits, already mentioned in Chapter 3.

Although we have worked for simplicity with an illustration involving only the most extreme kind of data, similar arguments could be used with, for instance, a family of six children made up of one boy and five girls. The relevant P value would be calculated by finding the probability of this particular pattern of events, together with the more extreme case of six girls, as well as the analogous patterns with boys and girls interchanged. To do this would only require knowledge of the binomial probabilities already discussed in Chapter 2.

The above example illustrates the essential characteristics of all significance tests, although there are naturally a number of variations in detail according to the frequency distributions involved and the exact form of null hypothesis chosen. Individual requirements for the most important tests will be given with the applications described later.

It is important to realise that a significance test is not an infallible guide. In fields of research involving a large degree of natural variability in the material studied, there are bound to be very considerable opportunities for making mistakes. The main purpose of a significance test is to provide a reasonably objective basis for reaching decisions. In general, if the probability P is very large or very small, then the test agrees with intuitive common sense in providing for the continued acceptance or rejection, respectively, of the null hypothesis. But when P lies in the broad intermediate range of values, say from 20 per cent down to 0.1 per cent, the use of a specific objective rule is invaluable.

An important aspect of this kind of test described is that the significance level usually determines the extent to which a certain type of error is made. Thus, if we work with a 5 per cent level then we shall reject the null hypothesis, when it is in fact true, on 5 per cent of the occasions that we use the test. Some workers may prefer to operate more cautiously, with, say, the 1 per cent level, reducing the rejection of a true null hypothesis to one occasion in a hundred. There is, however, a limit to the extent to which this caution can usefully be carried. We should soon find ourselves in the position of the sceptic who so rarely rejected any null hypothesis that he prevented himself from adopting a new theory even when this was unmistakably the right thing to do. As mentioned before, the 5 per cent level seems to be quite serviceable for general scientific use. The exact level used is often not very important. Unless there are special circumstances calling for a different value, we should normally always work with the same level so as to avoid unconscious bias or prejudice creeping into the interpretation of experimental results. If, however, we were considering experiments to test a new, and possibly dangerous drug, we should probably want to achieve a much smaller value of P, like 1 per cent or even 0.1 per cent, before recommending the new drug for general use.

There is, therefore, some advantage, when reporting the results of experiments, in giving as close an indication as possible of the actual value of P attained. Suppose, for example, that P is found to be less than 1 per cent. It is better to state this explicitly, rather than merely record that P is less than 5 per cent.

In the case of Mrs Smith's children, discussed above, the significance test was performed without any direct estimation of the chance of any child being a girl; we considered the assumption that the chance *was* one-half and calculated the consequences in practice. Sometimes, however, it is convenient to make a direct estimate of some quantity we are interested in, particularly if the samples of data are fairly large. Suppose in a genetical experiment we are examining the possibility of linkage between two loci, and suppose that we obtain an estimate of 41.2 per cent for the recombination fraction, based on a fairly large sample. Using the standard error of this estimate, say 2.8, we can calculate confidence limits as described in Chapter 3. The 95 per cent limits would be $41.2 \pm (1.96 \times 2.8) = 41.2 \pm 5.5$, i.e. 35.7 to 46.7. Now this range does not include 50 per cent, which is the value we expect if there is no linkage. Since 95 per cent limits exclude exactly 5 per cent of probability, we can say that there is a significance departure at the 5 per cent level of significance from the null hypothesis of no linkage (when the recombination fraction is exactly 50 per cent).

The result is exactly the same as that which would be obtained by following the rules suggested earlier in this chapter. The precise method of procedure is discussed in more detail in Chapter 5. The chief point we are making here is that if it is possible to assign confidence limits to an estimate, then there is a significant departure from any value suggested by a null hypothesis which does not fall between the limits; and the corresponding P is the probability excluded by limits which just fail to include the null-hypothetical value.

We can sum up this general discussion by saying that the fundamental purpose of a significance test is to provide a tool by means of which we can decide whether or not apparent differences in recorded data can be reasonably attributed to chance variations.

The skilful handling of significance tests comes only with practical experience, but there is one particular pitfall which must be avoided in drawing conclusions. It follows from the foregoing discussion that if the result of a test is significant, we can hardly ignore it with any justification: the null hypothesis requires modification at least. If, on the other hand, the result is non-significant, this does *not* automatically imply positive evidence in favour of the null hypothesis. The data considered might be so scanty as to contain very little information about the hypothesis, null or otherwise. If, however, the data are extensive, then non-significance may imply that appreciable departures from the null hypothesis are unlikely since these would usually lead to a significant result. In some cases the situation can be clarified by the use of confidence intervals. Normally, positive evidence in favour of a scientific theory results from a gradual build-up of circumstantial evidence, though this never amounts to 'proof' in the strict sense. In fairly exact sciences such as physics and chemistry, a single falsified crucial prediction demands at least some modification of existing theory. Similarly, in the biological sciences a statistically significant result is one that must normally be taken seriously.

5

Simple significance tests based on the normal distribution

Although in the previous chapter some simple examples of significance tests were given, we were mainly concerned with the underlying ideas rather than with numerical details. We must now go on to consider more carefully the routine calculations involved in some of the situations that arise most frequently in practice. In this chapter we shall look at a number of simple tests based on the normal distribution. As mentioned before, this distribution is of very wide applicability: sometimes it will represent sufficiently well the variation of the observed measurements themselves; sometimes it can be used as a convenient approximation to other distributions such as the binomial or Poisson.

5.1 COMPARISON WITH A KNOWN STANDARD

Let us suppose that we have captured a few birds of a particular species that appear to be living on a diet that is rather different from usual. Various questions can be asked about the possible effect of a different diet on physiological function. To illustrate this we shall merely take the problem of deciding whether there seems to be any real difference in body-weight from the known average for the species.

We shall suppose for the moment that the distribution of body-weight is known fairly accurately from a large number of observations and that it is approximately normal in character. We thus know both the true mean and standard deviation to quite a high degree of precision. Let these actually be 95.61 g and 7.82 g respectively.

Suppose we capture three birds whose average weight is 89.33 g and ask whether this is significantly different from the known average of 95.61 g. As explained in Chapter 3, the standard error of the mean of three values will be $7.82/\sqrt{3} = 4.52$.

If we take as null hypothesis the tentative view that the three captured birds are really representatives of the population regarded as the standard, then the difference (here -6.28) between the observed mean and the standard will have a normal distribution about *zero*, and the standard deviation of this distribution will be 4.52. The observed difference is only 1.39 ($=6.28/4.52$) standard deviations from zero, so this is evidently not a very extreme type of occurrence. We know that 5 per cent of observations lie in the tails of a normal distribution more than 1.96 standard deviations away from the mean. Thus, the ratio of any observed difference to its standard deviation which is less than 1.96 will be non-significant at the 5 per cent level. In the present case we should say that there was no evidence, at the 5 per cent level of significance, that the three captured birds showed anything more than ordinary chance variations in weight from the average for the species as a whole. However, if we had a sufficiently large number of measurements of birds feeding on the unusual diet, we could estimate the mean more accurately and some real departure from the standard might be established.

In general, we have a sample of n observations with mean $\bar{x}$ and standard deviation s. And we want to compare the observed $\bar{x}$ with a hypothetical value μ. If we do not know the true standard deviation σ, we must use the sample estimate s. We then calculate

$$d = \frac{\bar{x} - \mu}{s/\sqrt{n}}, \tag{13}$$

and treat d as a normal variable with zero mean and unit standard deviation, determining the significance level achieved by reference to Appendix 1. Confidence limits for the unknown population mean μ are

$$\bar{x} - ds/\sqrt{n} \quad \text{to} \quad \bar{x} + ds/\sqrt{n}, \tag{14}$$

where d is chosen from Appendix 1 to correspond to the required probability. These results are all right provided n is sufficiently large, say greater than 30. For smaller n the methods of sections 6.3 and 6.4 must be used.

If, however, we do know σ fairly accurately, as in the above example, we can use it instead of s in the expressions in formulae (13) and (14). The results are then valid even for small n, and in the illustration we had $n = 3$.

When the standard μ is not known to a high degree of accuracy compared with $\bar{x}$, we are really comparing two *samples*. We then resort to the methods of section 5.2 if the samples are large, or sections 6.3 and 6.4 if they are small.

5.2 COMPARISON OF MEANS OF TWO LARGE SAMPLES OF SIZES $n_1 > 30$, $n_2 > 30$

Let us now take a slight variation of the illustration used in the last section, and suppose that we have examined a large number of birds of the species in question and wish to determine whether there is any real difference in average weight between the two sexes. With reasonably accurate measurements there is almost certain to be some apparent difference. The question is whether it is likely to be due only to chance variation. We shall assume that the body-weights of each sex are approximately normally distributed; or, less stringently and more realistically, that the means of large numbers of such weights are so distributed.

The first thing to do is to calculate the mean and variance for each sex separately, as described in Chapter 2, using of course automatic methods for preference. Thus, we might have 125 male birds whose weights had mean 92.31 and variance 56.22, grammes being the units used. The variance is of course the variance of the individual measurements. To estimate the variance of the mean we require $56.22/125 = 0.4498$, as already explained in Chapter 3. Let the corresponding mean and variance for a series of 85 females be 88.84 and 65.41 respectively. This time the variance of the mean is $65.41/85 = 0.7695$. Now the difference between the two means is $92.31 - 88.84 = 3.47$. And the variance of this difference is the sum of the corresponding variances, i.e. $0.4498 + 0.7695 = 1.2193$. We are here using the standard result that the variance of the sum *or* difference of two *independent* variables is, in either case, the sum of the two individual variances. The standard deviation of the difference is therefore $\sqrt{1.219} = 1.104$. We now have all the ingredients required for carrying out a significance test.

The natural null hypothesis in this example is that the difference between the true means is zero. We assume that the means of large numbers of observations of either sex are normally distributed. It therefore follows that the difference in observed means (here found to be 3.47) would have a normal distribution about *zero*. Moreover, we have estimated the standard deviation of the latter distribution as 1.104, which is likely to be quite an accurate figure since the two samples are both large. The observed difference in means, 3.47, is 3.14 $(=3.47/1.104)$ standard deviations away from zero. This result is certainly significant at the 5 per cent level, corresponding to a ratio of 1.96, and in fact reference to Appendix 1 shows that the P value is slightly less than 0.2 per cent.

As explained before, this does not *prove* that there is a real difference. It means that we should provisionally accept the idea of a real difference rather than suppose that a very improbable type of event had occurred.

We can quote the observed difference, with its standard error, in weight between the two sexes as 3.47 ± 1.10 g. Thus, 95 per cent confidence limits are $3.47 \pm (1.10 \times 1.96) = 1.31$ to 5.63 g, and we can use this range of values in further discussions to give an indication of what the true difference might be. With more observations the range would be smaller.

General formulae

It is not difficult to apply the foregoing reasoning to any problem in which the general conditions are analogous. In order to summarise the technique, we shall exhibit it in terms of some simple algebra. Let us suppose that we have two groups of observations which are labelled '1' and '2', using these numbers as distinguishing suffixes. The first group contains n_1 observations having mean $\bar{x}_1$ and estimated variance s_1^2, calculated as indicated in Chapter 2; and the second group contains n_2 observations having mean $\bar{x}_2$ and variance s_2^2. The observed variances of the two means, $\bar{x}_1$ and $\bar{x}_2$, are therefore s_1^2/n_1 and s_2^2/n_2 respectively. We then calculate the ratio d given by

$$d = \frac{\bar{x}_1 - \bar{x}_2}{\sqrt{\left(\dfrac{s_1^2}{n_1} + \dfrac{s_2^2}{n_2}\right)}}. \tag{15}$$

Now, on the assumption that the individual values in each group are normally distributed, or that at any rate the means $\bar{x}_1$ and $\bar{x}_2$ are each normally distributed, it follows that d is also a normal variable with unit standard deviation. To test the hypothesis that the true mean of d is zero, we merely refer its value to Appendix 1.

As written, d may turn out to be positive or negative, according as $\bar{x}_1$ is greater or less than $\bar{x}_2$. If the absolute value of d, i.e. the actual numerical value, ignoring the sign, is greater than 1.96, then there is a significant difference between the true, but unknown, means (say μ_1 and μ_2) at the 5 per cent level of significance. Thus, if $d = -2.20$, the result would certainly be significant at the 5 per cent level.

It is also useful to calculate confidence limits for the difference, $\mu_1 - \mu_2$, between the two true means. These are

$$(\bar{x}_1 - \bar{x}_2) - d\sqrt{\left(\frac{s_1^2}{n_1} + \frac{s_2^2}{n_2}\right)} \quad \text{to} \quad (\bar{x}_1 - \bar{x}_2) + d\sqrt{\left(\frac{s_1^2}{n_1} + \frac{s_2^2}{n_2}\right)}, \tag{16}$$

where as usual d is chosen from Appendix 1 to correspond to the required probability.

5.3 THE 'NORMAL' APPROXIMATIONS TO BINOMIAL AND POISSON DISTRIBUTIONS

For strict accuracy, the test described in the last section requires that the means $\bar{x}_1$ and $\bar{x}_2$ shall be normally distributed. Nevertheless, the test is still extremely useful when dealing with measurements whose distributions are only approximately normal. It frequently happens that mean values tend to be nearly normal when they are based on large samples, even if the individual measurements are not so distributed. The formulae in sections 5.1 and 5.2 can therefore be applied with some confidence in quite a wide variety of circumstances. When a test based on the assumption of an underlying normal distribution turns out not to be very sensitive to departures from this assumption, we say that the test is *robust*.

So far in this chapter we have been thinking of specific measurements such as height and weight. It is, however, possible to apply the theory of the normal distribution to other situations where the basic distributions are something quite different, such as the binomial or Poisson. This application usually requires the samples to be large.

Comparing a percentage based on a large sample of size $n > 30$ with a known standard

First, let us consider the comparison of a percentage, based on a large sample involving a binomial distribution, with a known standard. Suppose, for example, that we have a genetical experiment designed to investigate the possibility of two loci being linked. We find 23 recombinants out of a total of 65 organisms examined. If there is no linkage, the true value of the recombination fraction will be $p = \frac{1}{2}$. The observed recombination fraction, on the other hand, is $23/65 = 0.354$. We can expect the number of recombinants to follow a binomial distribution, and, as the sample is a large one, we can also use the normal approximation.

As stated earlier in section 2.5, the mean and variance of the observed proportion are p and pq/n respectively. Moreover, on the null hypothesis $p = \frac{1}{2}$. The variance is thus $0.25/65 = 0.003846$, and the standard error is $\sqrt{0.003846} = 0.062$. The departure of the observed value 0.354 from the hypothetical 0.500 is 0.146, which is appreciably greater than twice the corresponding standard error. We thus conclude that there is significant evidence in favour of linkage at the 5 per cent level, i.e. $P < 0.05$.

Suppose, in general, we have a sample of n units, a of which exhibit some character, and wish to compare the observed proportion a/n with a hypothetical value p. We then require

$$d = \frac{(a/n) - p}{\sqrt{\left(\frac{p(1-p)}{n}\right)}}, \tag{17}$$

where d is taken to be a normal variable with zero mean and unit standard deviation.

If the result of the above test is significant, we shall probably want to set confidence limits for the true unknown proportion. This is done as already indicated in section 3.3.

Comparing two percentages based on two large samples

Suppose that we want to compare the germination rate of spinach seeds for two different methods of preparation, which we can call A and B. We have sown 80 seeds of the first type, of which 65 germinate; and 90 of the second type, with 80 germinating. It is assumed here that the actual sowing has been done in such a way that neither A nor B will be specially favoured. The elimination of this kind of bias is an important aspect of experimental design and will be referred to again later in Chapter 11. Providing that steps have been taken to avoid such bias we can validly compare the two germination rates, which are 81.2 per cent and 88.9 per cent respectively, as follows.

As before, we choose as the null hypothesis the suggestion that there is no real difference between the two germination rates, and then examine the observed variation between the two samples of data. We first require an estimate of the common germination rate. This is best made from the two samples taken together. Thus a total of 145 seeds germinated out of 170 planted, giving a rate of 85.3 per cent.

Now, as we can see from Chapter 2, the number of seeds germinating is likely to have a binomial distribution. But as the data are fairly abundant, we can use the normal approximation. We first find the variances of the two estimates. Working in decimals for convenience, the first germination rate of 0.812 has the variance

$$\frac{0.853 \times 0.147}{80} = 0.00157,$$

and the second germination rate of 0.889 has variance

$$\frac{0.853 \times 0.147}{90} = 0.00139.$$

The variance of the difference between the two estimates, i.e. $0.889 - 0.812 = 0.077$, is accordingly the sum of the two individual variances, i.e. 0.00296. The square root of the latter gives the standard deviation of the difference, which we find to be 0.054. The ratio of the difference to its standard deviation, i.e. $0.077/0.054 = 1.43$, is appreciably less than 1.96. So there is no question of a significant difference at the 5 per cent level.

It is important to notice that we are calculating the two variances from the germination rate estimated from the two samples bulked together. This is in accordance with the null hypothesis.

In general, if there are two samples, say n_1 units having a_1 with some character and n_2 having a_2 with the same character, we first find

$$k_1 = \frac{a_1}{n_1};\ k_2 = \frac{a_2}{n_2};\ k = \frac{a_1 + a_2}{n_1 + n_2}.$$

The quantity d, which we have used before as a normal variable with zero mean and unit standard deviation, is then

$$d = \frac{k_1 - k_2}{\sqrt{\left[k(1-k)\left(\frac{1}{n_1} + \frac{1}{n_2}\right)\right]}}, \tag{18}$$

and we refer to Appendix 1 as usual.

As mentioned above, the samples should be fairly large for this test to be satisfactory. We should also have k_1 and k_2 not too near to zero or unity.

Confidence limits for the difference $p_1 - p_2$ between the two unknown percentages are

$$(k_1 - k_2) - d\sqrt{\left[\frac{k_1(1-k_1)}{n_1} + \frac{k_2(1-k_2)}{n_2}\right]}$$

$$\text{to} \quad (k_1 - k_2) + d\sqrt{\left[\frac{k_1(1-k_1)}{n_1} + \frac{k_2(1-k_2)}{n_2}\right]}, \tag{18a}$$

where d is chosen from Appendix 1 to correspond to the required probability.

Comparing two Poisson distributions

It sometimes happens that the individuals which we observe are not part of a two-fold classification as in the last example. We simply record them if they exist. A good illustration of this is the number of bacterial colonies growing on a plate of culture medium. With a good uniform technique, we should expect to find that the number of colonies observed followed an approximately Poisson distribution. There is, in general, no question of classifying the original cells plated out into those that grew and those that did not.

If we compared two plates, one with 51 colonies and one with only 32, we might well ask whether the observed difference of 19 was likely to be due to chance. With supposedly identical conditions, a significant difference would point to some failure in technique. On the other hand, we might have deliberately introduced some variation in, say, the nutrients in the culture medium.

Since the mean of a Poisson distribution is also the best

estimate of its variance, we can estimate the variance of the difference 19 by adding the individual estimates of variance, i.e. $51 + 32 = 83$. This leads to a standard deviation of $\sqrt{83} = 9.11$. The ratio $19/9.11 = 2.09$ indicates that the difference between the two plates is quite definitely significant at the 5 per cent level.

Suppose, in the general case, that we have two samples of observations. The first has n_1 observations, *each* following the same Poisson distribution, with mean $\bar{x}_1$. The second has n_2 observations from a Poisson population with mean $\bar{x}_2$. We calculate

$$d = \frac{\bar{x}_1 - \bar{x}_2}{\sqrt{\left(\frac{\bar{x}_1}{n_1} + \frac{\bar{x}_2}{n_2}\right)}}, \tag{19}$$

and refer to Appendix 1 as before.

This test will work reasonably well provided $n_1\bar{x}_1$ and $n_2\bar{x}_2$ are both larger than about 30. Note that in the numerical illustration above we had $n_1 = n_2 = 1$.

Confidence limits for the unknown difference $m_1 - m_2$ are

$$(\bar{x}_1 - \bar{x}_2) - d\sqrt{\left(\frac{\bar{x}_1}{n_1} + \frac{\bar{x}_2}{n_2}\right)} \quad \text{to} \quad (\bar{x}_1 - \bar{x}_2) + d\sqrt{\left(\frac{\bar{x}_1}{n_1} + \frac{\bar{x}_2}{n_2}\right)}, \tag{19a}$$

where d is chosen from Appendix 1 to correspond to the required probability.

To compare two small single observations, we can refer to *Biometrika Tables for Statisticians*. Note that the relevant table is 'one-tailed'; see section 5.4 below.

5.4 ONE- AND TWO-TAILED TESTS

It is now time to mention an important, though slightly tricky, point in connection with significance tests that we have so far managed to avoid. In all the examples already discussed we have been concerned to decide whether certain observed differences were significant or not. We made no mention in the null hypotheses that we were interested only in differences in a particular direction. Thus, when examining the possibility of a sex-difference in weight for a certain species of bird, we were quite prepared to believe that males were on average really heavier than females, or vice versa. This being the case, it was entirely proper to take into account the probabilities in both tails of the relevant normal distribution. In other words, when considering the difference $\bar{x}_1 - \bar{x}_2$, large positive or large negative values would be equally interesting. Sometimes, however, this approach is inappropriate, as the following example shows.

Suppose we normally use a standard treatment for a serious infection of sheep and that the animals either die or make a good recovery. A new treatment is now suggested which is claimed by its sponsors to give a higher recovery rate. We are really interested only in evidence that implies the new treatment is significantly *better* than the standard. If the new treatment seemed to be less successful than the old, or if it seemed to be only doubtfully better, we should keep to the well-tried standard method. This is equivalent to taking notice of differences in recovery rates in one direction only; when considering the approximately normal distribution of the observed difference in recovery rates, we are interested in the probabilities in only one tail, the one corresponding to an apparent improvement.

It should be clear that the ratios previously used for given significance levels were relevant to 'two-tailed' or 'two-sided' tests, and that the present 'one-tailed' or 'one-sided' situation requires a different criterion. The adjustment is easy to make.

The normal curve is symmetrical, and a deviation of 1.960 standard deviations from the mean in either direction cuts off a total of 5 per cent in the two tails taken together, i.e. 2.5 per cent in each tail. If therefore we used a criterion of 1.960 standard deviations for one-tailed tests, the effective significance level would be 2.5 per cent and not 5 per cent. Obviously, the deviation for 10 per cent in the two-tailed case, namely 1.645, gives us 5 per cent in one tail. This ensures that in the long run the use of one-tailed tests, where appropriate, will still mean that the null hypothesis will be rejected when true only 5 per cent of the time.

A similar rule applies to any other significance level. If we want to use a 1 per cent level in a one-tailed test, then we must use deviations corresponding to the 2 per cent level for a two-tailed test.

Suppose, for example, we applied formula (18) to the problem of testing the new treatment for the sheep disease mentioned above. (We are assuming of course that the infected animals available for experimentation are allotted to the two treatments at random, so as to eliminate possible bias; see Chapter 11.) Let us use '1' to represent the standard treatment and '2' the new one. We further supppose that of n_1 animals given the standard treatment a_1 recover, while of n_2 animals given the new treatment a_2 recover. Thus k_1, k_2 and k are the observed recovery rates for the old treatment, the new treatment and the whole group of animals respectively. We should then claim significance at the 5 per cent level only if d turned out to be less than -1.645 (note the negative sign which follows from the way we have labelled the two treatments). If we were more reluctant to adopt a new and comparatively untried method (we might fear, for instance, the likelihood of undesirable long-term side-effects), we might prefer to use the 1 per cent level, for which the corresponding value of d would be -2.326.

6

The use of t-tests for small samples

6.1 THE IMPORTANCE OF SMALL SAMPLES

In Chapter 5 we have seen how to carry out a number of simple significance tests when the underlying distribution was either normal or at least approximately normal. It was, however, usually necessary to assume that samples were large, or alternatively that with small samples (as in the comparison of a few observations with a known standard) some additional information was available about the ideal population, such as accurate indications of its mean and variance. Unfortunately, information of this type is often lacking in practice, and we are faced with basing decisions on small samples of data which must be judged and interpreted largely on their own internal evidence.

Many people feel that small samples are so inherently unreliable that it is best not to draw conclusions from them at all. A little reflection will show that having this attitude will result in being unable to deal with a number of very common difficulties. In the first place, it is often not possible to increase the data to any appreciable extent: the number of animals with some rare disease that are available for experimentation may well be quite small, and we are obliged to make the best of it. Again, environmental conditions may be so variable in time and space that we can obtain reasonable homogeneity only within the compass of a comparatively small set-up. Also, whether we like it or not, much scientifically relevant evidence that is available from the work of ourselves or others is, in fact, rather scanty. It is a counsel of despair to ignore it. And in any case people will, inevitably, draw conclusions from rather inadequate material, so the most sensible procedure is to try to use methods of analysis that will tell us objectively what conclusions, however vague, can justifiably be drawn from the data. We have already seen that the large amount of variation in biological material necessitates the use of statistical methods even in large samples. Similar methods involving the same kind of probability interpretations can be adapted for use with small samples of data. It might be thought that even if such methods could be used, the amount of information extracted would be so small as to be hardly worth the effort. Fortunately, and perhaps surprisingly, this is not so. The utility of small sample tests is now widely recognised, and it is worth while acquiring the facility of carrying out the simplest and most frequently required tests for oneself. The remainder of this chapter deals with the so-called 't-test', originally developed by 'Student'. Other devices of this sort will appear later in the book.

6.2 COMPARISON OF SAMPLE MEAN WITH A STANDARD (VARIANCE UNKNOWN)

The simplest significance test in Chapter 5 was that involving the comparison of a sample of measurements with some known standard. If the standard entailed a normal distribution with *known* mean and variance, then the test worked even if the sample was small. However, it is very common in small sample applications, with say $n \leqslant 30$, for the true variance not to be known at all, even if it is legitimate to assign to the mean some hypothetical value such as zero. In this case the best we can do is to estimate the variance from the sample, and use the quantity t given by

$$t = \frac{\bar{x} - \mu}{s/\sqrt{n}}. \tag{20}$$

We are now using t, instead of the d given in formula (13), because we want to consider samples of any size.

The snag is that we no longer have a normal distribution, but a special new distribution called 'Student's' t-distribution. However, it is a remarkable fact that the sampling behaviour of t depends only on n, and is independent of the values of μ and σ. This is especially important here because it is precisely the case where σ is unknown with which we wish to deal. The distribution of t has been extensively tabulated for various values of n. When n is large t behaves very much like the normal variable d, which has zero mean and unit standard deviation. Appendix 2 gives a number of convenient significance points for the purpose of carrying out the resultant rests. These are conducted in a manner very similar to that already described in connection with the normally distributed deviation d. The example below should make the procedure clear.

Two special points should first be mentioned. The first is that the distribution now depends on the value of n, and so care must be taken to see that the table is entered on the correct line. A phrase often used in this connection is the *number of degrees of freedom*, which is equal to $n - 1$ in the present context. Some authors write n for the number of degrees of freedom, so that the sample size is then $n + 1$. It is important, therefore, in assessing other people's results in detail to establish which notation has been used. The second point is that the same distinction arises as previously with the normal distribution in connection with one-sided and two-sided tests. Appendix 2 applies immediately as it stands to two-sided tests; for one-sided tests we make the modifications previously described, namely entering the table at twice the level actually required.

We now consider an illustration of a t-test where there is essentially only one set of measurements to be tested. Suppose we want to compare the effects of two analgesic drugs A and B in the treatment of a certain disease, and have eight patients who are willing to co-operate in the experiment. (In some cases one of the 'drugs' might be merely an inert placebo used as a 'control' to test the other drug.) There is quite likely to be considerable variation amongst the patients in age, sex, severity of symptoms,

Table 4. *The effects of two analgesic drugs*

Patient	Hours of relief with analgesic A	Hours of relief with analgesic B	Relative advantage of of B in hours (x)
1	3.2	3.8	+0.6
2	1.6	1.0	−0.6
3	5.7	8.4	+2.7
4	2.8	3.6	+0.8
5	5.5	5.0	−0.5
6	1.2	3.5	+2.3
7	6.1	7.3	+1.2
8	2.9	4.8	+1.9
Mean	3.62	4.67	+1.05

general state of health, environmental conditions and so on. We therefore decide to try both drugs on each patient, recording the number of hours of freedom from pain. Questions of experimental design arise here which we shall consider in more detail in Chapter 11. For the moment we shall assume that reasonable precautions are taken, e.g. that sufficient time is allowed to elapse between the administrations of both drugs to avoid 'overlapping' effects; that in four patients A is given first and in four B is given first; that the patients are unaware of which drug is being administered; etc. Typical data are shown in Table 4.

Inspection of the table shows that on the whole drug B did better than A; in six cases there was an apparent improvement, sometimes substantial, while in two cases there were better results with A. Nevertheless, there is a considerable amount of variation present, patients 2 and 6 giving on average a rather low response to treatment, in contrast to 3 and 7 who respond well. It may be doubted whether we should be justified in concluding that B was better than A.

Because of the great variation between patients, it is better to consider directly the set of figures in column 4 of the table, rather than to compare the two sets of numbers in columns 2 and 3. The former procedure is called the *method of paired comparisons*. The average response of each patient is, of course, eliminated from the figure giving the relative advantage of drug B over drug A. Moreover, if we take as null hypothesis the assumption that A and B are equally effective, then the true average advantage of B will be zero, i.e. $\mu = 0$. We are therefore led to test whether the mean $\bar{x}$ is significantly different from zero. If the number of observations was large, we could calculate t in formula (20), and treat it as a normal variable with zero mean and unit standard deviation, since then the difference between s and σ would be unimportant. This type of test has already been discussed in Chapter 5. We have, however, only a small sample with $n = 8$, and therefore use a t-test. The main calculations are as follows.

$$\bar{x} = \frac{1}{n}\sum x = \frac{8.40}{8} = 1.050,$$

$$s^2 = \frac{1}{n-1}\left\{\sum x^2 - \frac{1}{n}\left(\sum x\right)^2\right\}$$

$$= \frac{19.24 - 8.82}{7} = \frac{10.42}{7} = 1.489,$$

$$\frac{s}{\sqrt{n}} = \frac{1.220}{2.828} = 0.4314,$$

$$t = \frac{\bar{x} - \mu}{s/\sqrt{n}} = \frac{1.050 - 0.000}{0.4314} = 2.434.$$

As described earlier, we can in practice readily calculate $\bar{x}$ and s by automatic methods. Reference to Appendix 2 shows that to achieve significance at the 5 per cent level with 7 degrees of freedom requires a t-value of 2.365 or more. There is accordingly no doubt that the data examined do involve a statistically significant advantage in drug B, with $P < 0.05$.

If drug B were some modification of A that would hardly be expected on pharmacological grounds to give worse results, we should have a 'one-sided' situation. We should then base the test on one tail of the distribution only – that for which t had large positive values. As with the normal distribution, we multiply the required one-sided probability level by two. To achieve a true 5 per cent level, we enter Appendix 2 in the 10 per cent column giving, for 7 degrees of freedom, the value of 1.895, which is considerably exceeded on the present data.

It is instructive to compare the rows in Appendix 2. The final row for an infinite number of degrees of freedom, when we have a normal distribution as the limiting form of t, is not very markedly different from the case when there are only 30 degrees of freedom. This justifies us in using the normal approximation for samples that have more than about 30 observations.

As with previous tests, the question arises as to what happens if there are appreciable departures from the basic assumptions. We saw that in large sample tests there are no serious difficulties provided that the means are approximately normally distributed. Moreover, there are good reasons for expecting this to occur even when the basic distributions are not normal. Similar considerations apply to the t-test, which fortunately turns out to be rather 'robust' (in the sense already referred to at the beginning of section 5.3) for quite appreciable departures from normality. The best plan seems to be to use the t-test wherever possible and applicable, but to do so with caution, especially if there is any reason to suspect that the underlying distributions are in some way peculiar.

6.3 COMPARISON OF MEANS OF TWO SMALL SAMPLES (UNKNOWN VARIANCES ASSUMED EQUAL)

We now come to the problem of comparing the means of two small samples. What we require is the small sample analogue of the large sample test described in section 5.2, and under the right conditions we can use an extension of the t-test introduced in the previous section. Apart from the basic requirement that the underlying distributions should be at least approximately normal, we shall for the time being make the further assumption that the true variances of the two populations to be compared are the same. What should be done when the variances are unequal is somewhat controversial, but the test mentioned below in section 6.5 is easily applied and is unlikely to give trouble.

Let us consider the type of problem previously discussed in

section 5.2 in relation to large samples. Assume that we now have two samples of birds, at least one of which is small. The first contains n_1 observations and has mean $\bar{x}_1$ and variance $s_1{}^2$, while in the second the corresponding quantities are n_2, $\bar{x}_2$ and $s_2{}^2$. The null hypothesis to be tested is that the means μ_1 and μ_2 are equal. The modified form of 'Student's' t is

$$t = \frac{\bar{x}_1 - \bar{x}_2}{s\sqrt{\left(\dfrac{1}{n_1} + \dfrac{1}{n_2}\right)}}, \tag{21}$$

where s is an estimate of the standard deviation based on both samples jointly. For technical reasons we must *not* pool all the data and then find the root-mean-square about the mean of all $n_1 + n_2$ observations, but instead proceed as follows. We calculate in the usual way the sums of squares about the mean for each sample *separately*; then add these two sums and divide by the total number of degrees of freedom involved, i.e. $n_1 - 1$ for the first sample and $n_2 - 1$ for the second, making $n_1 + n_2 - 2$ in all. We thus have

$$\begin{aligned} s^2 &= \frac{\sum_1(x - \bar{x}_1)^2 + \sum_2(x - \bar{x}_2)^2}{n_1 + n_2 - 2} \\ &= \frac{1}{n_1 + n_2 - 2}\left\{\sum_1 x^2 - \frac{\left(\sum_1 x\right)^2}{n_1} + \sum_2 x^2 - \frac{\left(\sum_2 x\right)^2}{n_2}\right\}, \end{aligned} \tag{22}$$

where $\sum_1$ means summation over the first sample only, and $\sum_2$ sumation over the second sample only. Although formula (22) looks a little complicated, it is not really so in practice; no new principle is involved, and the sums of squares about the means are formed as before.

All we have to do therefore is to calculate t from formula (21), using the value of s indicated by formula (22), and to refer to the distribution of t (e.g. Appendix 2) with $n_1 + n_2 - 2$ degrees of freedom.

Suppose that we had much smaller numbers of observations than those available in section 5.2, and that there were 10 male birds with mean 90.80 g and 9 females with mean 81.52 g. We shall assume that the sums of squares about the means have been found for each sample separately and are 497 and 530. This gives variances of $497/9 = 55.2$ and $530/8 = 66.2$, which are not very different from each other considering the sampling variation likely (the possibility of a statistical test of the equality of variances is mentioned below in section 6.5). Using formula (22) gives the estimate

$$s = \sqrt{\left(\frac{497 + 530}{9 + 8}\right)} = \sqrt{60.41} = 7.772.$$

The resultant value of t is therefore, from formula (21),

$$t = \frac{+9.28}{7.772\sqrt{(\frac{1}{10} + \frac{1}{9})}} = \frac{+9.28}{3.571} = +2.60.$$

Reference to Appendix 2 shows that for 17 degrees of freedom this value is just beyond the 2 per cent point. We are of course thinking here in terms of a two-tailed test. The usual remarks apply when the situation calls for a one-tailed test.

6.4 CONFIDENCE LIMITS

While we are dealing with t-tests, it should be pointed out that the t-distribution can be used to attach confidence limits to estimates in the same way that the normal distribution can be used. Thus, in the case of comparing a single sample with a standard, formula (20) leads to the confidence limits for the unknown population mean μ given by

$$\bar{x} - ts/\sqrt{n} \quad \text{to} \quad \bar{x} + ts/\sqrt{n}, \tag{23}$$

where the value of t to be inserted is whatever corresponds to the required degree of confidence on $n - 1$ degrees of freedom. Thus, with 5 degrees of freedom the 5 per cent point is 2.571. We therefore put $t = 2.571$ in formula (23) in order to derive a 95 per cent confidence interval.

The analogous result for the comparison of two samples, when we want a confidence interval for the difference between the two means, is obtained from formula (21) as

$$(\bar{x}_1 - \bar{x}_2) - ts\sqrt{\left(\frac{1}{n_1} + \frac{1}{n_2}\right)} \quad \text{to} \quad (\bar{x}_1 - \bar{x}_2) + ts\sqrt{\left(\frac{1}{n_1} + \frac{1}{n_2}\right)}. \tag{24}$$

In the worked example, comparing the mean weights of male and female birds we had $\bar{x}_1 - \bar{x}_2 = +9.28$. With 17 degrees of freedom the 5 per cent point is 2.110. Thus the required 95 per cent confidence interval is

$$+9.28 \pm (2.110 \times 3.571) = 1.75 \quad \text{to} \quad 16.81.$$

We can thus say that the odds are 19 to 1 on that the average male is heavier than the average female by an amount lying between 1.75 and 16.81 g. The significance test was, as we saw, in favour of a real non-zero difference. The confidence interval does not, therefore, contain the value zero, though it does in this example cover a rather wide range.

6.5 COMPARISON OF MEANS OF TWO SMALL SAMPLES (UNKNOWN VARIANCES NOT ASSUMED EQUAL)

In the above treatment of the problem of comparing the means of two samples we assumed that the true variances were in any case identical, although of course the sample values s_1^2 and s_2^2 would to some extent differ. The conscientious investigator would therefore make a point of *testing* statistically for the equality of the variances. This is done by the so-called variance-ratio or F-test. Detailed examples are given later in Chapter 11, where we discuss applications to the analysis of variance. F is defined as the ratio of the larger to the smaller of the two variance estimates s_1^2 and s_2^2, and so F is never less than unity. It is convenient to label the samples so that s_1^2 is larger than s_2^2; we thus have

$$F = \frac{s_1^2}{s_2^2}. \tag{25}$$

All we then have to do is to find from tables of the F-distribution the appropriate value of F for the chosen level of significance corresponding to $f_1 = n_1 - 1$ degrees of freedom in the numerator and $f_2 = n_2 - 1$ in the denominator. If this tabulated value is exceeded in the data, the result is significant.

A word of warning is necessary here. The most common use of the F-test is in the analysis of variance which requires only a one-tailed test. Tables are therefore normally presented in this form. In the present context, however, we are equally interested in deviations in both directions. To work at the 5 per cent level of significance here, we must use a table giving 2.5 per cent points. Appendix 5 gives only 5 per cent and 1 per cent points. More extensive tabulations are available in *Biometrika Tables for Statisticians*, which provides 10, 5, 2.5, 1, 0.5 and 0.1 percentage points.

It should also be mentioned that there is a certain lack of 'robustness' in the F-test, so caution should be exercised in the interpretation of results if appreciable departures from normality are suspected.

If we decide that the true variances of the two samples are probably the same, then we can carry on with the t-test described in section 6.4. If, on the other hand, a significant difference between s_1^2 and s_2^2 is found, a special procedure is required. There are in fact various ways of coping with the difficulty. One of the easiest to use is the following approximation. We return to formula (15) in section 5.2, namely

$$d = \frac{\bar{x}_1 - \bar{x}_2}{\sqrt{\left(\dfrac{s_1^2}{n_1} + \dfrac{s_2^2}{n_2}\right)}}, \qquad (15)$$

but now treat this as being distributed approximately like 'Student's' t with f degrees of freedom, the latter being given by

$$\left.\begin{aligned} f &= \frac{1}{\dfrac{u^2}{n_1 - 1} + \dfrac{(1-u)^2}{n_2 - 1}}, \\ \text{where} \quad & \\ u &= \frac{s_1^2/n_1}{s_1^2/n_1 + s_2^2/n_2}. \end{aligned}\right\} \qquad (26)$$

Confidence limits for the difference, $\mu_1 - \mu_2$, between the two sample means are given by formula (16), namely

$$(\bar{x}_1 - \bar{x}_2) - d\sqrt{\left(\frac{s_1^2}{n_1} + \frac{s_2^2}{n_2}\right)} \quad \text{to} \quad (\bar{x}_1 - \bar{x}_2) + d\sqrt{\left(\frac{s_1^2}{n_1} + \frac{s_2^2}{n_2}\right)}, \tag{16}$$

remembering that we are now treating d as a t with f degrees of freedom.

It should be realised that f, as calculated from formula (26), will not in general turn out to be an integer, as has been the case so far. However, it will usually be sufficiently accurate to interpolate linearly between the values of t shown in Appendix 2 corresponding to *adjacent* integral values of the degrees of freedom.

7

Contingency tables and χ^2

7.1 CONTINGENCY TABLES

Up to the present we have been studying the analysis used for investigating various quantitative measurements, such as stature, the number of bacterial colonies growing on a plate of culture medium, the germination rate of seeds, and so on. Many important problems are, however, not readily put into this form. We may, for instance, be able to classify hair colour very broadly into fair, brown, black, red and perhaps one or two intermediate shades, but not feel justified in trying to make a very precise assessment. The same kind of difficulty also arises with eye colour. Suppose we have a large number of individuals each of which can be classified in a more-or-less qualitative way for both hair and eye colour. What can be said, if anything, about the association between hair colour on the one hand and eye colour on the other?

Typical data can be set out in a two-way array like the main entries in Table 5, where we have 227 males classified for both hair and eye colour. It was not thought possible to distinguish at all accurately between grey eyes and green eyes, and so these shades were grouped together. Again, there were so few cases of

Table 5. *Contingency table relating hair colour and eye colour (observations are the main entries; expectations are in brackets)*

Eye colour	Hair colour			
	Red/fair	Brown	Black	Total
Blue	65 (44.5)	26 (36.2)	8 (18.3)	99
Grey/green	32 (43.6)	41 (35.5)	24 (17.9)	97
Brown	5 (13.9)	16 (11.3)	10 (5.7)	31
Total	102	83	42	227

red hair that it was decided to combine these with fair hair, so as to avoid having a column in the table with very small numbers and therefore likely to be very much affected by sampling variation.

It is a very common occurrence to have data which can be exhibited in this sort of pattern. Whenever we have two methods of cross-classification, each of which is made up of several more-or-less qualitative subdivisions, we can arrange the data in the form of a contingency table. Sometimes it is convenient, as we shall see in Chapter 8, to extend the technique to quantitative variables which are subjected to broad groupings, e.g. the division of stature into 0.02 m intervals, or even into just three classes such as 'short', 'medium' and 'tall'.

We now turn to the problem of assessing the material presented by a contingency table. An obvious common-sense way of making a preliminary investigation is to examine each row (or column) separately and express all the items in each row as percentages of the row total. If each row then exhibited much the same series of percentages, we should conclude that the classification involved in differences between rows had little influence on the classification giving rise to the columns. That is, we should decide that there was little or no *association* between the two classifications. This intuitive approach is in fact quite sound, but unless the number of individual units is very large there will be appreciable disturbances due to sampling variation, and the issue may be in doubt. The best way to carry out a reliable statistical test is to calculate a new quantity called χ^2 (pronounced chi-squared, with hard 'ch'), which is, amongst other things, a measure of the extent to which the observed numbers in the cells of a contingency table depart from the values we should have if the rows of percentages mentioned above were all identical. We then refer to tables of the χ^2 distribution to see whether the observed value is larger than would be expected by chance on a null hypothesis that postulated no association whatever between the two classifications.

The required calculations are, in general, best performed in stages (although there are in fact direct short-cut methods) so as to reveal the location of the disturbances, if any, that are present. We first find the number to be expected in the long run in each cell of the table, supposing that we could indulge in repeated sampling in which the marginal totals were kept fixed but the numbers in the body of the table could follow the sort of underlying distribution we should expect. The 'expected' number for any cell is given quite simply by dividing the product of the two corresponding marginal totals by the grand total. Thus the expected number of individuals in Table 5 with red/fair hair and blue eyes is $102 \times 99 \div 227 = 44.5$, which is appreciably different from the observed value of 65. This sort of calculation is repeated for each cell of the table, and the expected numbers for Table 5 are enclosed in brackets below the observed values.

Direct comparison of the observed and expected numbers gives an immediate clue to possible departures from independence. Thus, a big relative discrepancy can be discovered in Table 5 in the cell for black-haired individuals with blue eyes, the observations showing a considerable deficit. We must have, however, an objective numerical estimate of the importance of such departures. Now, whenever we can break down our data into a set of compartments for each of which we have both an observed (O) and an expected (E) number, the appropriate value of χ^2 is given by

$$\chi^2 = \sum \frac{(O - E)^2}{E}, \tag{27}$$

where the summation symbol $\sum$ indicates summation over all the compartments. It should perhaps be mentioned at this point that the quantity on the right-hand side of formula (27) is often referred to nowadays as X^2 (instead of χ^2), in order to distinguish between X^2, which is calculated from the observations, and the theoretical χ^2 distribution used for comparison. In the present elementary book we have not sought to maintain this distinction by using different symbols.

As with the t and F already considered, the concept of *degrees of freedom* is also crucial here. It is essential to know in any particular context how to assess accurately the number of degrees of freedom involved, and the appropriate rules will be given in the various applications described. For the moment it is sufficient to know that in the case of a contingency table the number of degrees of freedom is the product of one less than the number of rows and one less than the number of columns. Thus, in Table 5 there are $2 \times 2 = 4$ degrees of freedom. It is perhaps worth mentioning that this is, in fact, the number of cells in the body of the table that can be filled in *independently* if the margins are regarded as fixed.

If we perform the calculations implied by formula (27) we find

$$\chi^2 = \frac{(65.0 - 44.5)^2}{44.5} + \frac{(26.0 - 36.2)^2}{36.2} + \ldots$$

$$= 35.0, \text{ with 4 degrees of freedom.}$$

Some useful percentage points for different numbers of degrees of freedom are given in Appendix 3. We see there that with 4 degrees of freedom χ^2 must exceed only 18.5 for significance at the 0.1 per cent level. The data in Table 5 do therefore provide very strong evidence in favour of some kind of association between eye colour and hair colour, with $P < 0.001$.

We can summarise the above procedure algebraically as follows. The general contingency table with r rows and c columns can be written as in Table 6.

For the top left-hand entry we have $O = a$ and $E = AB/N$. Similarly, to obtain the expected value for any other entry, we multiply the corresponding marginal totals and divide by the

Table 6. *A general $r \times c$ contingency table*

a	b	...	A
c	d	...	.
⋮	⋮	⋮	
B	.	...	N

grand total. The appropriate value of χ^2, with $(r - 1)(c - 1)$ degrees of freedom, is then calculated from formula (27).

With large contingency tables the calculation of χ^2 can be lengthy, tedious and very liable to error. Such situations are best dealt with by a computerised approach, using special statistical software that can handle the data automatically.

The test just described is intended to be used when the numbers of observations are small enough for sampling variation to leave us in some doubt as to the interpretation of the data. Nevertheless, there is a lower limit to the size of sample for which the method is sufficiently reliable. The best rule to follow is that no *expected* number should be smaller than about 5, although the *observed* numbers may be less. If some cells do exhibit this undesirable feature, then we can sometimes overcome the difficulty by amalgamating certain rows or columns of the table. In fact, as mentioned in the illustration above, we had already done this in respect of red hair. If this colour had been given a separate column some very small expected numbers would have appeared.

When the association exhibited by a contingency table is significant, it is sometimes useful to have a direct measure of its strength. Although χ^2 provides a suitable significance test, it does not, as it stands, give an immediate measure of the degree of association. This can be seen from the fact that if we keep the relative proportions in the table all constant but increase the total number of observations, then χ^2 will be directly proportional to that total. A convenient measure of association is accordingly χ^2/n – the *mean square contingency* – where n is the total number of observations. This index is quite useful for comparing in a general way the associations involved in several contingency tables, where each has the same degrees of freedom but may contain very different numbers of observations. A further modification for comparing tables with different degrees of freedom is to divide the latter into the mean square contingency. For a table with r rows and c columns we should use $\chi^2/\{n(r-1)(c-1)\}$, an index which makes due allowance for variations in the numbers of both the observations available and the degrees of freedom.

7.2 SPECIAL CASE OF A TABLE WITH ONLY TWO ROWS

In the previous section we have seen how to deal with contingency tables of any size. If very many rows and columns are involved, the calculations are inevitably somewhat laborious. However, there are one or two special cases in which there is some advantage in trying to short-cut the work by the use of a more compact formula. We first consider the case of a table which has only two rows (or columns), although there may be several columns (or rows). Suppose, for instance, we are studying the sex-ratio of the offspring of different stallions. We are quite prepared to find departures from a 1:1 ratio, but are primarily interested in discovering whether the ratio varies from stallion to stallion more than could be expected by pure chance. Table 7 shows data on the offspring of six stallions.

The overall totals of males and females certainly show an excess of the former. We can easily establish this by finding the estimated proportion of males, $k = 71/116$, and testing whether this is significantly different from 0.5. One fairly quick method is to use the normal approximation to the binomial distribution and so compare k with the value 0.5, using the standard error $\surd(0.25/116)$, as explained in section 5.3.

If, however, we examine the individual columns we see that the first three stallions and possibly the fifth may show no more than ordinary variations from a 1:1 ratio, but that the fourth and sixth seem quite likely to have an appreciable excess of males. A test of homogeneity of the several sex-ratios is therefore of some importance. We can easily carry out the χ^2-test of the previous section. But the point to be made here is that the calculations are performed more readily as follows, especially when some form of

Table 7. *Numbers of males and females amongst the offspring of several stallions*

Sex of offspring	Reference no. of stallion 1	2	3	4	5	6	Total
Male	13	8	7	15	9	19	71
Female	9	10	8	6	5	7	45
Total	22	18	15	21	14	26	116

automatic calculation is available, by what is often called Brandt & Snedecor's formula. We compute

$$\chi^2 = \frac{\dfrac{9^2}{22} + \dfrac{10^2}{18} + \dfrac{8^2}{15} + \dfrac{6^2}{21} + \dfrac{5^2}{14} + \dfrac{7^2}{26} - \dfrac{45^2}{116}}{\dfrac{71}{116} \times \dfrac{45}{116}}.$$

It will be seen that we are simply summing the squares of each element in the 'female' row divided by the corresponding column total, deducting the analogous ratio for the column of totals and finally dividing by $k(1-k)$.

A useful check is provided by the fact that precisely the same answer must be given by carrying out the main operation on the upper row of the table. We actually find $\chi^2 = 6.03$, with 5 degrees of freedom. There is, therefore, no evidence of significant heterogeneity since the 5 per cent point is at 11.07.

The general formula for a table with two rows and c columns is easily remembered in the following terms. Let a typical member of the top row in any column be a, and that of the lower row be b, where $a + b = n$. Let the corresponding quantities in the 'totals' column be A, B and N, where $k = A/N$, $1 - k = B/N$. A typical array is shown in Table 8. Brandt & Snedecor's formula is then

$$\chi^2 = \frac{\sum \dfrac{a^2}{n} - \dfrac{A^2}{N}}{k(1-k)} = \frac{\sum \dfrac{b^2}{n} - \dfrac{B^2}{N}}{k(1-k)}, \tag{28}$$

Table 8. *A general 2 × c contingency table*

...	a	...	A
...	b	...	B
...	n	...	N

the degrees of freedom being one less than the number of columns, namely $c - 1$.

It will be realised that this type of contingency table analysis is effectively testing the homogeneity of a whole set of binomial distributions, and that the χ^2-test can be validly applied for quite small numbers so long as the expectation in each cell does not fall below about 5 units.

7.3 SPECIAL CASE OF 2 × 2 TABLES

A further simplification is possible if we consider contingency tables which involve only two rows and two columns, usually called 2×2 tables. These are very common in practice, and a quick method of analysis is highly desirable. We are, in essence, comparing two percentages and, with sufficiently large numbers, could use the procedure of section 5.3. But the method described below is in many respects more serviceable. Not only is it quick in practice, but we can easily decide whether or not the numbers are too small for a valid application, i.e. when any expectation is less than 5. Let us consider the 2×2 contingency table whose elements are represented by the symbols in Table 9.

If we follow through algebraically the rules recommended in the previous section, we shall finally arrive at a χ^2 given by

$$\chi^2 = \frac{n(ad - bc)^2}{(a + b)(c + d)(a + c)(b + d)}, \tag{29}$$

where there is now only *one* degree of freedom. The characteristic features of the right-hand side of formula (29) are that in the numerator we have the grand total n and the square of the difference of cross products ad and bc, while the denominator is merely the product of all marginal totals.

Table 9. *A general 2 × 2 contingency table*

Second classification	First classification		
	Presence of A	Absence of A	Total
Presence of *B*	a	b	$a + b$
Absence of *B*	c	d	$c + d$
Total	$a + c$	$b + d$	n

We must still be careful about having no expected number less than about 5 (this can usually be checked by inspection and approximate mental arithmetic). Nevertheless, there is a useful correction due to Yates that is worth applying if some of the numbers are on the small side. In fact, it is probably always best to use the correction as a matter of routine, as it can do no harm when numbers are large. The need for a correction arises from the fact that the theory of χ^2 assumes that the numbers involved in a contingency table are really continuously variable and not discrete. It is possible to see that a useful way to offset this is to replace a, b, c and d by a set of numbers which are 0.5 of a unit less extreme, i.e. exhibit whatever association may be present to a slightly less degree. Fortunately, we do not have to worry about how to decide this point, which some people find tricky, as a general formula will automatically make the right adjustment. We use in fact

$$\chi^2 = \frac{n\{|ad - bc| - \frac{1}{2}n\}^2}{(a + b)(c + d)(a + c)(b + d)}, \tag{30}$$

where the vertical lines in $|ad - bc|$ mean that we are to take the absolute, i.e. positive, value of the difference between ad and bc. This difference is therefore always reduced in size by subtracting the correction $\frac{1}{2}n$, and the squared quantity in the numerator of formula (30) is less than the corresponding quantity in the numerator of formula (29). Formula (30) presents little difficulty in calculation, especially when using a pocket calculator.

Let us now consider a numerical example. Data on the effectiveness of inoculation against cholera are set out in Table 10.

We have two experimental groups of individuals. One of these groups has received the inoculation; the other has not. This involves the problem of obtaining an adequate 'control' group for the treatment under test. More will be said about this later when questions of experimental design are discussed. On the null hypothesis we suppose that the chance of infection is the same for all individuals, irrespective of whether they have been inoculated or not. As can be seen from the table, 21 per cent of the uninoculated were attacked, but only 11 per cent of the inoculated. Formula (30) yields the result

$$\chi^2 = \frac{200\{|869 - 1869| - 100\}^2}{100 \times 100 \times 32 \times 168} = 3.01.$$

Now, Appendix 3 gives 3.84 as the 5 per cent point when χ^2 has one degree of freedom, and so it appears as though no significant difference can be detected. However, an important point of interpretation arises here in connection with the distinction we have already discussed between one- and two-tailed tests. The analysis of a general contingency table introduced in section 7.1 is designed to detect any kind of departure from strict independence. In the present case of a 2×2 table, therefore, the figures in Appendix 3 as it stands relate to departures in 'either direction'. That is, they relate to discovering whether inoculation gives protection against cholera or whether it predisposes to attack (a not unknown circumstance with faultily prepared sera). This is the typical two-tailed situation. If we are testing for the existence of both kinds of effect, the data of Table 10 would not be significant at the 5 per cent level. If, on the other hand, we are confident before the trial that the inoculation could do no harm, i.e. will be at worst only useless, then we must test for departures in only one direction. The usual rule for reading one-tailed points from a two-tailed table then applies. For a true 5 per cent level we require the 10 per cent point shown in Appendix 3, namely 2.71 for one degree of freedom. In such a situation therefore a significant degree of protection would be indicated by the data.

Table 10. *The effectiveness of inoculation against cholera*

	Attacked	Unattacked	Total
Inoculated	11	89	100
Uninoculated	21	79	100
Total	32	168	200

7.4 EXACT TEST FOR 2 × 2 TABLES

In using the χ^2-test described above for examining contingency tables for departures from independence, we saw that it was unsafe to proceed if any expectation was less than about 5 units. With a large table it may be possible to amalgamate certain rows or columns in order to comply with this condition. But with only a 2 × 2 table, no such expedient is possible. The way out of the difficulty is to employ a test that is exact and not merely an approximation. This will be demonstrated in this section. The need for a test of this sort arises most frequently when the total number of individuals examined is small, but it can of course occur in samples of any size.

To obtain a clear idea of what is involved, let us suppose that we are comparing the effect of two treatments, A and B, on a serious but rare disease. We have in all only 9 cases and of 5 allotted to treatment A only 1 recovers, while all 4 patients given treatment B recover. What can be said about the relative merits of the two treatments having regard to the very small numbers involved? (We are assuming that the patients have been allotted in some suitable random fashion to the two treatments, e.g. *alternate* patients in order of occurrence to a given treatment.)

The best way to understand the usual test employed in this situation is to consider the basic approach as introduced and discussed in Chapter 4. The central idea is to take the null hypothesis that there is no difference between the treatments, and then consider the experimental results that might be observed. The actual result is judged significant if it belongs to a sufficiently extreme class of events. We have also seen above how contingency tables can be handled by examining the variation that can occur in the body of the table if the marginal totals are kept fixed. These lines of thought lead us to look in detail at the series of tables that might be observed. The data actually observed can be set out as in Table 11.

It is obvious on inspection that this is a relatively extreme type of table, since the bulk of the data falls in the top left and bottom right cells of the main part of the table. The whole series of possible results, for marginal totals the same as observed, can be represented in brief as follows:

4	1	3	2	2	3	1	4	0	5
0	4	1	3	2	2	3	1	4	0

Probabilities: $\frac{5}{126}$ $\frac{40}{126}$ $\frac{60}{126}$ $\frac{20}{126}$ $\frac{1}{126}$

The order of these arrays is important and results from concentrating attention on one cell (in this case we might have chosen the bottom left-hand entry for patients dying under treatment B) and making this cell take all possible values in succession, the rest of the table being filled in automatically. The table observed is at the left-hand extremity of the series of possible samples. A decision as to significance therefore rests on calculating the probability of occurrence of this table, and perhaps of an equally or more extreme table at the other end of the series.

Consider the array represented by the observations of Table 9, i.e.

a	b	$a+b$
c	d	$c+d$
$a+c$	$b+d$	n

The appropriate probability is in fact

$$\frac{(a+b)!(c+d)!(a+c)!(b+d)!}{n!a!b!c!d!}, \tag{31}$$

Table 11. *Results achieved in treating a rare disease with two different drugs*

	Died	Recovered	Total
Treatment *A*	4	1	5
Treatment *B*	0	4	4
Total	4	5	9

where the factorial $x!$ means $1 \cdot 2 \cdot 3 \ldots (x-1)x$, and 0! and 1! are both taken to have the value unity. Straightforward calculation then gives the probabilities shown below each array. On the extreme left we have for instance

$$\frac{5!4!4!5!}{9!4!1!0!4!} = \frac{5!5!}{9!} = \frac{5}{126}.$$

We must now decide whether we are to operate a one- or two-tailed test. If treatment B could hardly be expected to be worse than A we are only concerned with one tail, for which the relevant probability is $P = \frac{5}{126} = 3.97$ per cent, giving a significant result at the 5 per cent level. (In this example there are no more extreme tables to be taken into account.)

If, on the other hand, we are simply testing two treatments and wonder whether they might be different, then both tails must be taken into consideration. It is important to realise that the distribution in the kind of test under discussion is not necessarily symmetrical, so we must beware of merely doubling the probability in one tail. In the present two-tailed case the rule that we must add the probabilities for the observed result and any other more or equally extreme results involves the two arrays at either end of the series shown. The total probability is thus $p = \frac{6}{126} = 4.76$ per cent, which is still significant at the 5 per cent level, though only just. If we had doubled the probability in the tail examined first, we should have obtained the erroneous and non-significant figure of $2 \times 3.97 = 7.94$ per cent.

Since in the above illustration all the numbers are fairly small, there is little difficulty in evaluating formula (31). But there are occasions when we must use the exact test because one or more of the expectations in the body of the table is small, although some of the observed numbers, especially perhaps marginal values or the grand total, are quite large. Trouble arises because the factorials resulting from large observations may be enormous: thus 25! is a number with 26 figures in it! The solution of this difficulty using simple pencil and paper methods is to use logarithms, and the logarithms of all factorials up to 150! are conveniently given in Fisher & Yates' *Statistical Tables*. (*Biometrika Tables for Statisticians* is even more extensive and goes up to log 1000!) We simply add the logs of the numbers in the numerator of formula (31) and subtract the logs of the numbers in the denominator. The antilog of the result gives the required probability.

Consider, for example, the array

1	12	13
5	9	14
6	21	27

Formula (31) requires the evaluation of

$$\frac{13!14!6!21!}{27!1!12!5!9!}.$$

The logarithm of this to base 10 is easily found to be $-1.055\,898\,9 = \bar{2}.944\,101\,1$. Taking the antilog then gives the required probability, namely 0.0879 or 8.79 per cent. Similar calculations are performed for any other equally, or more, extreme arrays that must be taken into account in the significance test.

When using a pocket calculator that has a 'factorial' facility to calculate $x!$ directly (with some restriction on the size of x) the work is very much easier. But it is usually wise to alternate divisions and multiplications of factorials to avoid arriving at excessively large or small numbers that the calculator cannot handle. Thus in the above example we start with 13!, divide by 27!, multiply by 14!, divide by 12! etc. (ignoring 1! which makes no difference). We then quickly arrive at the same figure as that given above, i.e. 8.79 per cent.

7.5 SOME FALLACIES IN INTERPRETING CONTINGENCY TABLES

Difficulties are, of course, liable to arise in the interpretation of any kind of statistical analysis, especially when there is the possibility of some hidden bias or distortion being introduced. In this section we shall notice a special snag that arises with contingency tables.

Provided that the observational units are reasonably homogeneous, it is usually fairly safe to accept the results of the significance tests at face value. Thus, if we had a series of patients, all similar in sex, age group, severity of symptoms, etc., there would be little trouble in designing an experimental trial to test two treatments. Indeed, even if the patients were rather dissimilar, we could still obtain reliable results by allotting them strictly at random to the two treatments.

The real trouble arises when we unwittingly combine data from several heterogeneous sources into a single table. We might, for example, be looking for an association between certain diseases such as cancer or peptic ulcer and blood groups. The incidences of the diseases and of the blood groups may easily vary considerably in different parts of the country. This entails an essential heterogeneity in the data, and it would probably be unsafe to use contingency tables pooling data for the whole country. The answer is to construct a number of tables, each relevant to a fairly restricted geographical area or social group, and to test these separately.

In order to illustrate how spurious results may arise, let us consider a simple artificial example. Suppose the 2×2 tables for two distinct sources of data are

10	90	100
10	90	100
20	180	200

and

10	10	20
90	90	180
100	100	200

It is clear that in each table taken by itself there is no association whatever. Amalgamation, however, gives

20	100	120
100	180	280
120	280	400

which is highly significant as it stands, and evidently misleading.

The converse situation can also arise: an apparently non-significant table may be composed of significant components which cancel each other out when pooled. Thus

100	100	200
100	100	200
200	200	400

might be made up of

20	80	100
80	20	100
100	100	200

and

80	20	100
20	80	100
100	100	200

If these possible sources of error are clearly appreciated, it is usually not difficult in practice to examine one's data for hidden heterogeneity, and to make the proper allowance for it.

8

χ^2-tests of goodness-of-fit and homogeneity

8.1 INTRODUCTION TO GENERAL IDEA

In Chapter 7 the χ^2-test was introduced as a means of examining contingency tables for the existence of an association between two systems of classification. The null hypothesis was that the classifications were independent, and on this assumption the expected number of observations in each cell of the table could be calculated from the marginal frequencies. We then used the χ^2 given by formula (27) to test the agreement between hypothesis and observation. Precisely the same kind of test can be applied more generally, wherever in fact we want to examine a hypothesis which specifies the frequency with which the observations should fall into certain classificatory groups. This is usually referred to as a *goodness-of-fit* test. Some of the simplest applications are concerned with discovering whether a particular distribution, such as the normal or Poisson, is a sufficiently accurate description of the data. In other cases the hypotheses tested may be more elaborate. The use of formula (27), repeated below for convenience, is quite general. We have

$$\chi^2 = \sum \frac{(O - E)^2}{E}, \tag{27}$$

where O and E are the observed and expected numbers in any group, respectively, and $\sum$ indicates summation over all groups.

In the more general applications envisaged here special care is required to make sure that the correct number of degrees of freedom is assigned. It should be clear from the illustrations given later in this chapter just how this is done in practice. For the moment we merely note the general principles involved. To begin with, if a fixed number of observations is classified into a series containing k groups, then there are $k - 1$ degrees of freedom involved, since only $k - 1$ groups could be assigned arbitrary numbers of observations. Put another way, there would be k degrees of freedom if all groups could be filled arbitrarily, but 1 degree is lost because of the *constraint* introduced by the requirement that the total number of observations must be regarded as fixed.

This idea can be extended to additional constraints. It leads, for instance, to the rule already stated, that a contingency table with r rows and c columns yields $(r - 1)(c - 1)$ degrees of freedom. We can see that this is correct by considering the following argument. There are rc groups altogether. But we introduce $r + c$ constraints by regarding the marginal totals as fixed. However, the constraints themselves are not all independent since the row totals must have the same sum as the column totals. There are therefore only $r + c - 1$ independent constraints. The resultant number of degrees of freedom is thus $rc - (r + c - 1) = (r - 1)(c - 1)$.

In general, every independent constraint that is introduced absorbs one degree of freedom. Apart from the questions of rows and columns, etc., already mentioned, a common source of restriction is the estimation of parameters from the data. Suppose, for example, we are examining the hypothesis that the data can be described by a normal distribution. We are obliged to fit some specific normal distribution, and the best one is the curve for which μ and σ are estimated by the calculations described in Chapter 2. The estimation of the two parameters entails the loss of a further two degrees of freedom from the total otherwise available.

8.2 TESTING THE FIT OF A WHOLE FREQUENCY DISTRIBUTION TO DATA

In Chapter 2 the Poisson distribution was introduced in relation to the distribution of yeast cells in a haemocytometer. 'Student's' data was set out in Table 3. For any given number of cells per square we had both the actual observed number and the expected number. The latter was calculated from the Poisson probability given by formula (3) multiplied by the total number of squares. An important step in this procedure was the estimation of the Poisson parameter m from the mean of the sample $\bar{x}$. The observed and expected numbers which we wish to compare are shown in the second and third columns of Table 3. As already stated, the general agreement between the data and the hypothetical values appears to be quite good. What we now propose to do is to make a statistical test of this agreement, to test in a word the 'goodness-of-fit' of the Poisson distribution to the data.

In the discussion of the χ^2-test for contingency tables, it was pointed out that to be on the safe side we should try to arrange that no expectations were less than about 5 units. This rule should also be used in the more general applications now under discussion. Let us examine Table 3 with this requirement in mind. It is clear from the right-hand column that the first group for $x = 0$ and the last three groups for x greater than or equal to 11 are all below the acceptable minimum. The easiest course is to amalgamate the classes $x = 0$ and $x = 1$, and to pool the four classes for x equal to 10 or more. These modifications lead to the figures shown in Table 12.

We can now proceed with the calculation of χ^2 according to formula (27), obtaining

$$\chi^2 = \frac{(20.0 - 21.1)^2}{21.1} + \frac{(43.0 - 40.6)^2}{40.6} + \ldots + \frac{(9.0 - 8.7)^2}{8.7}$$

$$= 4.39.$$

Now we can calculate the degrees of freedom. The total of 400 squares in Table 12 is distributed amongst 10 classes. We therefore start off with 9 degrees of freedom. However, we must not

Table 12. *Distribution of yeast cells in a haemocytometer ('Student's' data) – condensed from Table 3*

Number of cells in a square (x)	Observed nos. of squares	Expected nos. of squares
0 or 1	20	21.2
2	43	40.6
3	53	63.4
4	86	74.2
5	70	69.4
6	54	54.2
7	37	36.2
8	18	21.1
9	10	11.0
10 or more	9	8.7
Total	400	400

forget the constraint introduced by the fact that we were obliged to estimate from the data the Poisson parameter m in order to calculate the expectations in the third column of the table. One further degree must therefore be deducted from the 9 originally available, leaving a total of 8. The 5 per cent point for χ^2 with 8 degrees of freedom is 15.5, so that there is no question of a significant departure from the hypothesis under test. It is clear that the Poisson distribution does afford a very satisfactory description of the data.

It is instructive to consider the corresponding problem presented by Table 2, in which the hypothetical distribution is binomial. Clearly, the expected number of families with two albinotic children is too small to stand on its own. If we amalgamate families with one and two albinos the expectation is 5.25. We now have only two classes to start with, and may well ask whether we shall end up with any degrees of freedom at all with which to make a test. The answer in this case is 'yes'. From the two classes we have a single degree of freedom. Moreover, there are no further constraints involved, since the hypothesis under test itself specifies the value of p in the binomial distribution; Mendelian theory suggests $p = \frac{1}{4}$, and no question of estimating p

from the data arises. We can therefore compute a χ^2 with 1 degree of freedom for testing the goodness-of-fit. It is obvious without calculation that the value so obtained would be insignificant. If, however, we were uncertain about the segregation ratio to be expected and were obliged to estimate it, this would absorb the single degree of freedom available, and no test of fit would be possible.

Suppose now that all entries in Table 2 were multiplied by 10. The third group could stand on its own, and so we would start with 2 degrees of freedom. A test of fit would then still be possible even if p were estimated from the data.

Similar methods can be used with any other well-specified distribution. When distributions are discrete, as are the Poisson or binomial, we are unlikely to run into serious difficulties because the individual observational classes automatically provide suitable groupings. With continuous distributions, for instance the normal, the picture is slightly different. Some suitable interval must be chosen for grouping, e.g. the 0.02 m intervals of Table 1. The observed numbers are immediately available: e.g. 14 for 1.74 m. To obtain the corresponding expected numbers, we have to find the *area* under the smooth curve. In the case of the normal distribution most of the work can actually be done by reference to published tables of the normal curve, or by direct computation using a suitable calculator. It is beyond the scope of the present book to discuss this in detail, but it may be of interest to indicate briefly the sort of calculation that may be involved.

The mean and standard deviation of the distribution in Table 1 were shown in Chapter 2 to be 1.703 m and 0.052 m respectively. What, for example, is the expected number for the 1.74 m class? When the group interval is not too large compared with the standard deviation, we can legitimately approximate to the area under a small part of the curve by the height of the curve multiplied by the width of the interval involved. Let us work in units of one standard deviation. The group interval is then $0.02/0.052 = 0.385$. The 1.74 m class is $(1.740 - 1.703)/0.052 = 0.712$ standard deviations from the mean. Using Fisher & Yates' *Statistical Tables*, we find that the ordinate corresponding to 0.712 is 0.310. The required area is then approximately $0.310 \times$

0.385 = 0.119, and the expected number is 117 × 0.119 = 13.9, which is in this case very close to the observed value of 14.

Alternatively, we can use a pocket calculator to evaluate formula (1) with $x = 1.740$, $\mu = 1.703$ and $\sigma = 0.052$, giving $y = 5.954$. Multiplying this by the group interval 0.02 and the sample size 117 then yields 13.9, as before.

If this argument were pursued in detail, we should probably find that we had to group together the first two classes and the last three. We should also remember that two degrees of freedom would be lost because both the mean and standard deviation have been estimated from the data.

8.3 TESTS OF HOMOGENEITY

Very often the data available to us fall naturally into several distinct groups. Experiments may have been carried out on different days or by different individuals, or may have been subject to some other source of possible variation in the environmental conditions. It is important for us to know whether these groups are effectively homogeneous in the patterns of statistical variation that they display or whether they are really heterogeneous. In the former case, we can pool the data before further analysis, but in the latter case it will probably be necessary to investigate each group separately.

Suppose, for example, that we want to estimate the gene frequencies for the ABO blood-group system in a previously uninvestigated country. We might hope to use pooled data for the whole country, simply recording the number of people in each of the groups A, B, AB and O. But on reflection we might, from past experience in other areas, suspect the existence of geographical variation. We should then try to obtain records of the numbers in each group for several separate areas. This then gives us a straightforward contingency table of the type already discussed in Chapter 7: one classification consists of the four blood groups, the other of the different areas. If χ^2 turns out not to be significant, we can pool the data for all areas; otherwise we must proceed by analysing each area separately.

Similar methods apply when we want to compare distributions

rather than the frequencies of various qualitative classes, as above. We have seen in section 8.2 how to test the fit of specific distributions to the available data. Earlier, in Chapters 5 and 6, methods were given of comparing two distributions such as the normal, binomial or Poisson by examining mean values. A more searching analysis, which will permit the comparison of several groups, can be made as follows provided sufficient data are available.

We might have several parallel haemocytometer counts of the kind exhibited in Table 3. With a good repeatable laboratory technique, which avoided systematic errors though not of course eliminating the usual statistical variations, the several distributions obtained should be homogeneous. This can be tested by making up a contingency table in which one classification involves the parallel counts and the other consists of suitable groupings of the observed numbers. A significant value of χ^2 would show heterogeneity between the different counts.

Various conclusions might follow from such an investigation. If the individual counts were adequately fitted by Poisson distributions, but there was heterogeneity between counts, we would suspect some systematic error in technique which altered the average level of the counts. If, on the other hand, the Poisson distribution failed to fit, we might look for some failure in technique which meant that the cells were distributed in the haemocytometer in some non-random way. This might or might not be combined with heterogeneity.

Several continuous distributions, e.g. the normal, can also be compared merely by using a suitable grouping of the measurements so as to yield a contingency table. It is important to realise that this general method of comparison makes no special assumptions about the actual form of the underlying distribution. We are merely testing the null hypothesis that the several distributions, no matter whether discrete or continuous, are really the same.

8.4 SMALL SAMPLES FROM THE NORMAL, BINOMIAL AND POISSON DISTRIBUTIONS

In the last section it was recommended that the distributions arising in parallel sets of data should be tested for homogeneity by suitably grouping the observations and building up a contingency table. When the total number of observations is small, it may not be possible to follow this advice without breaking the rule about expectations not being less than about 5. It is therefore worth drawing attention to one or two ways in which this difficulty can be overcome.

The normal distribution

When we have two small samples each supposedly drawn from a normal distribution, the tests described in Chapter 6 can be used for comparing the means and variances. In the present context these tests would be tests of homogeneity. If there are more than two samples, the matter is a little more complicated. However, the means of several groups, each having the same variance, can be tested for homogeneity by the 'analysis of variance' technique discussed later in Chapter 11; the homogeneity of the variances can be examined by Bartlett's test mentioned near the end of the same chapter in section 11.4.

The binomial distribution

In the binomial data of Table 2 we have exhibited the observed and expected frequencies for the whole distribution. With families of two children, there are three classes in all; with families of k, there would be $k + 1$ classes. Nevertheless, we can condense the table into just *two* columns by distinguishing only between normals and albinos, and this would of course apply for families of any size. Similar considerations could apply to any other data resulting from binomial distributions. In Table 2 there are 17 normals and 7 albinos. If we had further sets of data drawn from different sources, we could again build up a contingency table for testing homogeneity. This time, however, we could tolerate relatively smaller samples. In the special case where there are just two groups to be compared, it does not matter how small the numbers are, because we have a 2×2 contingency table which can be tested if necessary by the *exact* test of section 7.4.

The Poisson distribution

Small samples of Poisson data require a different treatment since the observations do not immediately lead to the kind of 'reduced' contingency table available with the binomial. At the end of section 5.3 it was shown how two Poissons might be compared using the normal approximation. What test, we may ask, is available to deal with more than two samples?

Suppose we have k parallel plates of culture media bearing bacterial colonies, and suppose that the number of colonies on any plate is represented by x. Then a suitable test of homogeneity is given by the χ^2

$$\chi^2 = \frac{\sum(x-\bar{x})^2}{\bar{x}} = \frac{1}{\bar{x}}\left\{\sum x^2 - \frac{1}{k}\left(\sum x\right)^2\right\}, \qquad (32)$$

where there are $k-1$ degrees of freedom. This test will work quite satisfactorily provided that the expected number for each plate, $\bar{x}$, is larger than about 5 units. It does not matter how small k is, although the test will not be very sensitive for low values. When $k = 2$, formula (32) reduces to the rule previously given at the end of section 5.3.

8.5 THE ADDITIVE PROPERTY OF χ^2 AND THE NORMAL APPROXIMATION

Although for practical purposes it is unnecessary to know much of the mathematical theory of the χ^2 distribution, there are two rather simple properties that can be very useful in applied work. The first of these is the so-called 'additive property' of χ^2. This means that if we add together the values of χ^2 calculated from different sources, then the resultant total is itself distributed like a χ^2, for which the number of degrees of freedom is the sum of the degrees of freedom arising from each individual source of data. Quite often the use of this result leads to the formation of a χ^2 with more degrees of freedom than are catered for in most available tables. This snag is easily overcome by making use of the second property, which is that if we have a χ^2 with f degrees of freedom, where f is fairly large, then

$$d = \sqrt{(2\chi^2)} - \sqrt{(2f - 1)} \tag{33}$$

is distributed approximately like a normal variable with zero mean and unit standard deviation. Significance can therefore be decided by reference to percentage points for the normal distribution, e.g. Appendix 1, remembering to follow the rules for a one-sided test.

Let us consider as a specific example a situation that could arise with data of the type shown in Table 12. We are using a haemocytometer to count, say, yeast cells, and are inclined to suspect some fault in the technique. Under satisfactory conditions, we have good grounds for expecting a Poisson distribution, and decide to test this hypothesis statistically. A special trial is made with counts being carried out for four different dilutions of the original suspension.

The type of analysis described above in section 8.2 is carried out on each of the four observed frequency distributions separately. We cannot, of course, pool the data because the dilutions are different, and so we know in advance that the theoretical Poisson distributions will not be the same. Table 13 shows the goodness-of-fit χ^2 obtained for each dilution, together with the associated number of degrees of freedom.

Now, the average value of a χ^2 is actually equal to the number of degrees of freedom on which it is based. All four dilutions in Table 13 therefore show χ^2 values well above expectation, though reference to Appendix 3 reveals the fact that none of these values is individually near the appropriate 5 per cent point. While we may suspect a tendency to heterogeneity, there is insufficient evidence to demonstrate this as the table stands. This is where we can use the additive property of χ^2 with advantage. The total χ^2 is 71.36 with 49 degrees of freedom. As we are beyond the scope of the tables of χ^2, we use the normal approximation indicated by formula (33). With $\chi^2 = 71.36$ and $f = 49$, we calculate d to be $\sqrt{142.72} - \sqrt{97} = 2.10$. This is appreciably greater than the ordinary 5 per cent point for a two-tailed test; the true significance level in the present case, where we must use a one-tailed criterion, is thus less than 2.5 per cent.

Table 13. *Amalgamation of homogeneity χ^2 from different samples*

Dilution	χ^2	Degrees of freedom
A	17.32	13
B	14.71	10
C	24.45	17
D	14.88	9
Total	71.36	49

Although none of the χ^2 values obtained from each dilution separately shows a significant departure from the Poisson distribution, amalgamation of the values gives a quite definite result. There seems to be little doubt that the technique of preparation is failing at some point. Moreover, all four dilutions seem to be affected in the same sort of way.

8.6 THE MEANING OF VERY SMALL χ^2 VALUES

In both this chapter and the previous one we have examined several applications of the χ^2 test. Each time the hall-mark of significance has been a sufficiently *large* value of χ^2. Small values can of course occur, though never negative ones. If the rows of a contingency table all happen to have exactly the same sets of percentages, when the latter are based on row totals, we shall actually obtain a χ^2 of zero. In fact, any exact agreement between observed and hypothetical frequencies will yield a zero goodness-of-fit χ^2. Such values should occur but rarely, and normally have no special importance attached to them. If, however, small values seem in any context to be too frequent, then we may have to give the matter special consideration, particularly if we can envisage a possible mechanism to explain the result.

A good example of how this peculiarity can arise is in the making of differential white-cell counts. Laboratory technicians are sometimes told that the sign of a good technique is to be able to obtain very similar results on repeated counts. Now we know that if repeated counts are made on a given blood film with a good technique the proportion of, say, polymorphonuclear cells observed should follow a binomial distribution. If a technician learns the lesson about repeatable results too well, we may find that the variation between counts drops considerably below what is expected. The usual explanation of this phenomenon (which may also appear in many other situations in different guises) is that unconscious bias influences the judgment so that all counts after the first tend to be systematically biased towards the original one. Deliberate cheating is of course possible, though not, so far as one can discover, the most common cause of this type of error.

The existence of the bias just described can be established by setting out the repeated counts for a given technician on a single blood film in the form of a contingency table, and calculating χ^2 as already described. We are now interested in *significantly small* values. This means that we are looking for values that are smaller than, say, the 95 or 99 per cent points corresponding to 5 or 1 per cent significance levels at the *lower* end of the distribution. A χ^2 that was less than 2.73 on 8 degrees of freedom would thus be regarded as evidence in favour of this kind of bias. The elimination of bias may be a more difficult matter than its detection. In the example just given, we can use mechanical counters operated in such a way that the technician does not know the number of cells in each group until after a count has been completed. Whatever device we decide to try, the data obtained can be tested by means of χ^2; small values indicating unconscious bias in recording and large values suggesting irregularities in preparation technique.

9

The correlation of measurements

9.1 THE GENERAL NOTION OF CORRELATION

A great many important numerical investigations are concerned with the association between two different kinds of measurement or classification. Several examples were given in Chapter 7 in connection with the use of contingency tables. These were primarily introduced to deal with a two-way classification of qualitative characters such as hair colour or eye colour, but could be applied to continuous variables if these were suitably grouped. Although indices such as the mean square contingency can be employed to measure the strength of the association present, as opposed to the use of χ^2 for a significance test, they are not ideal in situations where both characters show continuous variation of a kind that frequently arises in practice.

Consider, for example, the data in Table 14 in the next section, relating to the stature of fathers and sons. Each unit of observation is a father–son *pair*, and we can label the first measurement x and the second y. The table shows how a particular sample of family pairs was actually distributed. It is interesting to note that, as we should expect, tall fathers tend to have tall sons, and short fathers tend to have short sons. The association is by no means complete, although there is a preponderance of entries in the top left-hand and the bottom right-hand parts of the table. If the association were complete or nearly so, we should expect the entries to be concentrated along a diagonal set of cells. Alternatively, with no association the distribution of every row would be the same, and there would be no steady shift in row-means as we moved from row to row.

A popular index of the degree of association between continuous variables is the *correlation coefficient*. This is often applied in a rather loose and uncritical way to any data involving double classification of the type under discussion. Although such methods may be quite useful at times, they are not to be recommended. The reason is that the interpretation and distribution of the correlation coefficient are clearly understood only in certain rather special circumstances, and unless these are approximately fulfilled there may be considerable doubt about the statistical significance of any results obtained. This is a very important point, which is too often overlooked. However, it often happens that when *correlation* methods are inapplicable, we can use instead the *regression* techniques described in the next chapter. The present chapter and the next should therefore be studied in conjunction with each other. We now examine the conditions under which a correlation analysis is valid.

If we look at the marginal totals in Table 14, we have the distributions of fathers' heights and sons' heights taken independently of one another. Plotting either of these distributions in the form of a histogram (as described in Chapter 2) shows that we have something rather like the 'normal' form, which we know from experience occurs quite commonly with this kind of measurement. Inspection of the body of the table also shows a certain characteristic form. There is a concentration of frequency around the centre of the table with a falling off in all directions. This falling off occurs more rapidly in some directions than in others. (Actually, if we drew curves through points for which the true frequencies were constant we should obtain a set of concentric ellipses.)

In Chapter 2 we saw that the normal distribution is often a fair representation of particular kinds of data, and also that there were theoretical reasons for expecting this distribution to turn up in practice. Similarly, when there are two variables, as in Table 14, we often find that what is called a *bivariate normal distribution* is a reasonable description of the data. And again there are theoretical reasons for expecting such a distribution to appear, at least approximately. The mathematically curious may like to know the abstract definition of the distribution. Since there are now two basic variables, x and y, the analogue of the normal *curve* given in (1) is a bivariate normal *surface*. This is represented by

$$z = \frac{1}{2\pi\sigma_x\sigma_y\sqrt{(1-\rho^2)}} \exp\left[-\frac{1}{2(1-\rho^2)}\left\{\frac{(x-\mu_x)^2}{\sigma_x{}^2} - \frac{2\rho(x-\mu_x)(y-\mu_y)}{\sigma_x\sigma_y} + \frac{(y-\mu_y)^2}{\sigma_y{}^2}\right\}\right]. \quad (34)$$

Each variable considered separately has a normal distribution: x has mean μ_x and standard deviation σ_x, while the corresponding quantities for y are μ_y and σ_y. Inspection of formula (34) reveals an additional symbol ρ, which is the *correlation coefficient* between x and y. As with formula (1), no direct use will be made of the algebraic expression in formula (34). The calculations to which it leads can be defined quite simply in terms of the sums, sums of squares and sums of products of x and y, as explained below.

The correlation coefficient ρ is the required index of association between variables x and y. It can vary only within certain restricted limits, namely -1 and $+1$. When $\rho = 0$ there is no correlation, and each row of the table would have the same normal distribution. Similarly, all columns would have identical distributions (though not as a rule the same as the rows). When $\rho = +1$ we say that the correlation is complete or perfect. We can show that there must then be some *fixed* relation between the x and y of any pair of values, for example $y = ax + b$, where a and b are constants, and a is also positive. On the other hand, when $\rho = -1$ there is a complete negative correlation, in which there is again a fixed relation between each x and y, but this time a is negative. Thus whenever $\rho = -1$ or $+1$ there is no room for variation in, say, y when x is known. For the other values of ρ, the situation is intermediate between a fixed relation and complete independence. Whether the actual association present is positive or negative depends of course on the sign of ρ. It is intuitively clear from Table 14 that the value of ρ for these data must be some positive number between 0 and 1; there is obviously some positive association, but it falls far short of a perfect correlation for which all the entries would lie theoretically along a single line.

The correlation coefficient ρ is thus a characteristic of the ideal population of values from which actual observation pairs (x, y) are specific instances. We are already familiar with the idea of calculating from samples of observations estimates of ideal population means and variances. The same thing is true for the correlation coefficient. As shown in the next section, a quantity r, which is a suitable estimate of the true, but usually unknown, correlation coefficient ρ, can be calculated from any sample. In repeated sampling from the same population r will have a statistical distribution, in the same way that $\bar{x}$ or s have distributions in the case of a single variable considered earlier. This leads to significance tests for deciding whether any observed r is significantly different from zero, or perhaps from some other value of theoretical importance. Again we may wish to test whether two observed estimates differ significantly from each other. We shall now discuss the computational aspects.

9.2 THE CALCULATION OF AN ESTIMATED CORRELATION COEFFICIENT

In earlier chapters, where only a single normal distribution was involved, the basic problem was to calculate from the observations, represented by a set of x-values, estimates of the unknown mean μ and variance σ^2. This could be done using only the sample size n, the sum $\sum x$ and the sum of squares $\sum x^2$. The present situation entails pairs of observations, which we can represent by sets of associated values of x and y. From data consisting of n such pairs we want to estimate the two means μ_x and μ_y, the two variances $\sigma_x{}^2$ and $\sigma_y{}^2$, and the correlation coefficient ρ. This is in fact hardly more difficult than dealing with the single variable case. We need of course the sums $\sum x$ and $\sum y$, and the sums of squares $\sum x^2$ and $\sum y^2$; but we must have in addition the sum of products $\sum xy$. The latter expression means simply that we multiply the x and y occurring in each observation pair and sum over the whole sample. We now require not only the sums of squares about the means, $\sum(x - \bar{x})^2$ and $\sum(y - \bar{y})^2$, but also the sum of products about the means, $\sum(x - \bar{x})(y - \bar{y})$. There are several convenient ways of computing the latter, as we shall see.

When carrying out this sort of work it is as well to proceed systematically, so we first consider computing the basic sums, sums of squares and sums of products. As in section 2.5, we can easily accumulate sums such as $\sum x$ and $\sum x^2$ automatically. We can also obtain $\sum y$, $\sum y^2$ and $\sum xy$ just as easily. The basic quantities calculated from the data are therefore

$$\left.\begin{array}{lll} n & & \\ \sum x & \sum y & \\ \sum x^2 & \sum y^2 & \sum xy \end{array}\right\} \qquad (35)$$

and it may be found advantageous actually to set out the numbers like this in practice.

To find the sums of squares and products about the means, we use the obvious analogues of formula (7), namely

$$\left.\begin{aligned} \sum(x - \bar{x})^2 &= \sum x^2 - \frac{1}{n}\left(\sum x\right)^2, \\ \sum(y - \bar{y})^2 &= \sum y^2 - \frac{1}{n}\left(\sum y\right)^2, \\ \text{and} \quad \sum(x - \bar{x})(y - \bar{y}) &= \sum xy - \frac{1}{n}\left(\sum x\right)\left(\sum y\right). \end{aligned}\right\} \qquad (36)$$

The last equation in formula (36) is a new form, but its affinity with the other two is evident. We can now extend the array in formula (35) by adding a row of *correction factors* as shown in the first line of formula (37) below. Subtracting each of these from the number above then gives the required expressions on the left of formula (36). Thus, we have the additional rows

$$\left.\begin{array}{lll} \dfrac{1}{n}\left(\sum x\right)^2, & \dfrac{1}{n}\left(\sum y\right)^2, & \dfrac{1}{n}\left(\sum x\right)\left(\sum y\right), \\ \sum(x-\bar{x})^2, & \sum(y-\bar{y})^2, & \sum(x-\bar{x})(y-\bar{y}), \end{array}\right\} \quad (37)$$

where the quantities in the first line of formula (37) are subtracted from the corresponding quantities in the last line of formula (35) to give the second line of formula (37).

Dividing the last line of formula (37) by $n - 1$ gives the three quantities s_x^2, s_y^2 and c, where the first two are the already familiar estimates of variance as in formula (5). The third is a new quantity called the *covariance*, which is a kind of 'average product about the means', defined by

$$c = \frac{1}{n-1}\sum(x-\bar{x})(y-\bar{y}). \quad (38)$$

Note that the divisor is $n - 1$, as for variances. We can now calculate the estimated correlation coefficient by either of the two expressions given by

$$\left.\begin{aligned} r &= \frac{c}{s_x s_y} \\ &= \frac{\sum(x-\bar{x})(y-y)}{\sqrt{\left(\sum(x-\bar{x})^2\sum(y-\bar{y})^2\right)}}. \end{aligned}\right\} \quad (39)$$

With large samples a fair amount of labour might be involved in using only elementary calculations, though practice and systematic work would considerably reduce the time required. One method of making the computations more manageable is to use a certain amount of grouping, as indeed has already been done in Table 14. We may then wish to adjust the two variances by the corresponding Sheppard's corrections to give the best estimate of the correlation coefficient. The covariance, on the other hand, requires no grouping correction. However, it is usual to omit the corrections when proceeding to a significance test; to avoid unreliable results we should use groups that are reasonably narrow.

Let us consider the data of Table 14 by way of illustration. It is convenient here to take working origins for both x and y at 1.70 m, using working units of 0.02 m (i.e. 2 cm). The sums of $\sum x$ and $\sum y$, and the sums of squares $\sum x^2$ and $\sum y^2$, are easily obtained in the usual way from the marginal distributions. A modified method of finding $\sum xy$ is, however, preferable. To each cell of the table there correspond specific values of x and y. The product of these two values, with due regard to sign, has been entered in each cell in brackets. In order to find $\sum xy$, we must therefore multiply each of these products by the number of observations in the cell, i.e. the number of times that particular product occurs, and then sum over the whole table. Doing this for each quadrant of the table separately gives

$$\begin{array}{c|c} +684 & -115 \\ \hline -246 & +1447 \end{array},$$

so that $\sum xy = +1770$.

The whole set of calculations can therefore be displayed as follows:

$$n = 493 \qquad \bar{x} = 0.18 \qquad \bar{y} = 0.60$$

$$\sum x = 91 \qquad \sum y = 296$$

$$\sum x^2 = 3513 \qquad \sum y^2 = 3646 \qquad \sum xy = +1770$$

$$\frac{1}{n}\left(\sum x\right)^2 = 16.8 \quad \frac{1}{n}\left(\sum y\right)^2 = 177.7 \quad \frac{1}{n}\left(\sum x\right)\left(\sum y\right) = 54.6$$

$$\sum(x - \bar{x})^2 = 3496.6 \qquad \sum(y - \bar{y})^2 = 3468.3 \qquad \sum(x - \bar{x})(y - \bar{y}) = +1715.4$$

$$s_x{}^2 = 7.106 \qquad s_y{}^2 = 7.049 \qquad c = +3.487$$

$$s_x = 2.666 \qquad s_y = 2.655 \qquad r = +0.493$$

Table 14. *Joint distribution of stature*
are the products of

			Father's					
			1.58	1.60	1.62	1.64	1.66	1.68
		Working units	−6	−5	−4	−3	−2	−1
Son's stature in metres (y)	1.58	−6					1 (+ 12)	
	1.60	−5			1 (+ 20)	2 (+ 15)	2 (+ 10)	3 (+ 5)
	1.62	−4		2 (+ 20)	2 (+ 16)	3 (+ 12)	3 (+ 8)	5 (+ 4)
	1.64	−3		2 (+ 15)	4 (+ 12)	6 (+ 9)	5 (+ 6)	4 (+ 3)
	1.66	−2	1 (+ 12)	3 (+ 10)	3 (+ 8)	6 (+ 6)	9 (+ 4)	8 (+ 2)
	1.68	−1	2 (+ 6)	4 (+ 5)	5 (+ 4)	8 (+ 3)	9 (+ 2)	13 (+ 1)
	1.70	0		2 (0)	3 (0)	6 (0)	7 (0)	9 (0)
	1.72	1		1 (− 5)	2 (− 4)	5 (− 3)	8 (− 2)	8 (− 1)
	1.74	2			1 (− 8)	4 (− 6)	5 (− 4)	6 (− 2)
	1.76	3			1 (− 12)	1 (− 9)	5 (− 6)	4 (− 3)
	1.78	4				1 (− 12)	2 (− 8)	3 (− 4)
	1.80	5					1 (− 10)	1 (− 5)
	1.82	6					1 (− 12)	
	1.84	7						
	1.86	8						
Total			3	14	22	42	58	64

of fathers and sons (figures shown in brackets xy *when in working units)*

stature in metres (x)								
1.70	1.72	1.74	1.76	1.78	1.80	1.82	1.84	
0	1	2	3	4	5	6	7	Total
1 (0)								2
1 (0)								9
2 (0)	1 (− 4)	1 (− 8)	1 (− 12)					20
3 (0)	2 (− 3)		1 (− 9)					27
7 (0)	4 (− 2)	1 (− 4)	1 (− 6)		1 (− 10)			44
9 (0)	5 (− 1)	6 (− 2)	3 (− 3)	4 (− 4)		1 (− 6)		69
17 (0)	13 (0)	11 (0)	4 (0)	5 (0)	2 (0)	1 (0)		80
11 (0)	14 (+ 1)	9 (+ 2)	6 (+ 3)	2 (+ 4)	2 (+ 5)	1 (+ 6)		69
9 (0)	11 (+ 2)	8 (+ 4)	6 (+ 6)	3 (+ 8)	2 (+ 10)	1 (+ 12)	1 (+ 14)	57
4 (0)	9 (+ 3)	11 (+ 6)	3 (+ 9)	3 (+ 12)	1 (+ 15)	2 (+ 18)		44
4 (0)	4 (+ 4)	5 (+ 8)	6 (+ 12)	3 (+ 16)	1 (+ 20)	2 (+ 24)	1 (+ 28)	32
2 (0)	2 (+ 5)	3 (+ 10)	4 (+ 15)	4 (+ 20)	2 (+ 25)	3 (+ 30)	1 (+ 35)	23
	2 (+ 6)	1 (+ 12)	3 (+ 18)	1 (+ 24)	1 (+ 30)			9
		1 (+ 14)	2 (+ 21)	1 (+ 28)	1 (+ 35)			5
				1 (+ 32)	1 (+ 40)		1 (+ 56)	3
70	67	57	40	27	14	11	4	493

If we want to use Sheppard's corrections (see formula (8) on p. 21) we subtract $\frac{1}{12}h^2$ from the variance, h being the group interval. In working units $h = 1$, so we subtract 0.083.

Thus with Sheppard's corrections we have:

$$s'_x{}^2 = 7.023 \quad s'_y{}^2 = 6.966$$

$$s'_x = 2.650 \quad s'_y = 2.639.$$

The value $r = +0.493$ (or $+0.499$ if Sheppard's corrections are used) is very close to $+0.5$, which is what is expected on genetical grounds for this type of data. The question of attaching standard errors to such estimates and performing tests of significance will be discussed in the next section.

It should be noted, however, that pocket calculators are available with built-in statistical facilities that mean that it is necessary only to enter the set of (x, y) data pairs, with due regard to the repeated values in each cell. Pressing the relevant key then gives the correlation coefficient r automatically.

9.3 SIGNIFICANCE TESTS FOR CORRELATION COEFFICIENTS

Although there are various ways of interpreting the meaning of a correlation coefficient in absolute terms, we are usually more interested in whether the observed value is significantly different from zero, or whether two observed values are significantly different from each other. However, in some genetical applications concerning tests of multifactorial inheritance, certain non-zero values, like $\frac{1}{2}$ and $1/\sqrt{2}$, etc., have a special theoretical importance.

Normal approximation in large samples

If samples are sufficiently large, the estimate r tends to have a normal distribution with mean ρ and standard deviation $(1 - \rho^2)/\sqrt{n}$. We can then attach a standard error to the estimate and write

$$r \pm \frac{1 - r^2}{\sqrt{n}}. \tag{40}$$

Unless the true value ρ is very near to zero, the approach to normality is slow. As a general rule, it is probably best not to use formula (40) unless n is at least 500. In the example of the previous section n was not far short of this figure. Application of formula (40) leads to the result 0.493 ± 0.034. The difference 0.007 between the observed value of 0.493 and the hypothetical value of 0.500 is thus very small compared with its standard error. The observed value of the correlation coefficient in this sample is therefore not significantly different from 0.5.

Use of t-test for exact treatment when $\rho = 0$

When the sample size is not very large we may be faced with a more difficult problem, as the distribution of r is liable to have a rather awkward form. The special case $\rho = 0$, however, is easily dealt with. This arises in the situation when we are anxious to test whether the observed value is significantly different from zero, i.e. whether the hypothesis of independence is contradicted by the data. To carry out the required test we first calculate r and derive from it the quantity t given by

$$t = \frac{r\sqrt{(n-2)}}{\sqrt{(1-r^2)}}, \tag{41}$$

which is precisely a 'Student's' t with $n - 2$ degrees of freedom, and can therefore be referred to the usual tables, e.g. Appendix 2. To avoid the labour of actually calculating the right-hand side of formula (41) we can refer directly to Appendix 4, which gives the values of r that must be exceeded for significance at various levels according to the number of degrees of freedom available. If we had a sample containing 16 observation-pairs, i.e. with 14 degrees of freedom, then only values of r that were greater than $+0.497$ or less than -0.497 would be significant at the 5 per cent level.

A general test using the z-transformation

The changeover in the last paragraph from r, which is troublesome, to t, which is easily handled, is a good example of the use of a mathematical transformation in statistical work. This procedure is quite legitimate and respectable, as we are effectively carrying out the test we want to apply to the original variable, but are doing so by means of a more convenient but indirect mathematical method.

Troublesome distributions often turn up in statistics, and it is standard practice to look for a transformation that will turn the basic variable into one that is at least approximately normal. Conclusions about the latter are easily drawn. The presence or absence of significance in the transformed variable applies immediately to the original variable as well. Confidence limits for the new variable would, of course, need to be transformed back to the original terms.

These ideas are extremely useful in dealing with the estimate of a correlation coefficient when the hypothetical value is not zero, or when comparing two estimates of a coefficient not assumed to be zero. In general, the distribution of r is awkward. We therefore consider the transformed quantities

$$\left.\begin{aligned} z &= \tfrac{1}{2}\log_e \frac{1+r}{1-r}, \\ \text{and} \qquad \zeta &= \tfrac{1}{2}\log_e \frac{1+\rho}{1-\rho}. \end{aligned}\right\} \qquad (42)$$

The difference $z - \zeta$ is then approximately normally distributed with zero mean and variance $1/(n-3)$.

Let us use this more accurate method to compare the estimate of 0.493 derived from Table 14 with the hypothetical value 0.5. We find that

$$z = \tfrac{1}{2}\log_e \frac{1+0.493}{1-0.493} = 0.5400,$$

and

$$\zeta = \tfrac{1}{2}\log_e \frac{1 + 0.5}{1 - 0.5} = 0.5493,$$

so that

$$z - \zeta = -0.0093.$$

The standard deviation is $1/\sqrt{490} = 0.0452$. We now apply the ordinary normal significance test; since the departure of $z - \zeta$ from the theoretical value zero is still much less than the standard deviation, the result is certainly not significant.

A similar argument can be used to compare two correlation coefficients. The test required is now based on the fact that if z_1 and z_2 are the two transformed numbers, then $z_1 - z_2$ is approximately normal with mean $\zeta_1 - \zeta_2$ and variance $1/(n_1 - 3) + 1/(n_2 - 3)$, where the two samples contain n_1 and n_2 observations respectively.

Suppose we have obtained $r_1 = 0.523$ on a sample of $n_1 = 42$ and $r_2 = 0.836$ on a sample of $n_2 = 57$. We might, for example, have calculated the correlation coefficient between tail length and wing length in a certain species of bird for the two sexes separately, and wish to know whether there is likely to be a real difference in the degree of association present. We easily obtain $z_1 = 0.5804$ and $z_2 = 1.2077$. The difference is therefore 0.6273 with a standard error of $\sqrt{(\frac{1}{39} + \frac{1}{54})} = 0.2101$. Since the difference is practically three times its standard error, the result is strongly significant, with P not much larger than 0.2 per cent. Confidence limits for the difference $\rho_1 - \rho_2$ are obtained by first calculating these for $\zeta_1 - \zeta_2$ and then transforming back.

The transformation indicated in formula (42) is of course very easily handled when using a pocket calculator that supplies natural logarithms (that is, to base e) automatically.

9.4 GENERAL COMMENTS

The correlation coefficient is an extremely valuable statistical tool for detecting associations between continuous variables, but one should be careful to use it only when it is appropriate. As pointed out in section 9.1, we really need to be able to postulate a normal bivariate distribution for the pair of measurements made. If we collect a random series of animals and make two kinds of measurement on each, then there is good reason to hope that the requirement will be approximately fulfilled. *Both* measurements will certainly have distributions. Moreover, if the histograms of each kind of measurement taken separately look more or less as though the underlying distributions are normal, it may be reasonably safe to embark on the use of a correlation coefficient. An additional check is to look carefully at the two-way array given by the pairs of measurements. The distribution observed in each row and column ought to be a sample of values from some normal population. Any marked discrepancy would discourage the use of a correlation coefficient. (We are assuming here that elaborate statistical *tests* of bivariate normality are usually too time-consuming to be practicable.)

However, it is not uncommon in practice to find even these very loose conditions not being fulfilled. Suppose, to take an extreme example, we had a group of short men and a group of tall men. A graph or table plotting weight against height could certainly reveal the strong association existing between the two measurements. But the calculation of a correlation coefficient would not be valid because the heights would not be following a normal distribution, and there would be no possibility of bivariate normality.

In fact, it is more common than not to find the latter situation arising: one measurement follows some irregular, or perhaps arbitrarily chosen, pattern, while the other can quite legitimately be imagined to have a statistical distribution. This state of affairs can often be dealt with quite easily by the *regression* type of analysis which is described in the following chapter.

Although in the present chapter we have discussed correlation in terms of only one pair of variables, the general idea can be extended to any number of variables. We then talk about *partial correlations*. This important development will be examined in some detail in Chapter 14.

Another vital matter which must be mentioned here is the question of interpreting the meaning of a correlation coefficient.

As we have seen, independence, and complete association in a positive or negative sense, are represented by the numbers 0, +1 and −1, with intermediate numbers indicating intermediate states. What we must be careful *not* to assume is that because factors A and B are to some extent correlated, then A is at least partially responsible for B, or vice versa. This is fairly obvious if we consider two measurements such as height or weight, where both are directly related to growth in general but neither can be said to be the immediate cause of the other. Although a strong association may happen to be quite fortuitous, it is usually an indication that *some* connecting process is at work. Provided that we use correlation coefficients merely as pointers in this sort of way, we shall find that they are useful without being misleading.

10

Regression analysis

10.1 THE BASIC IDEA OF REGRESSION

In the previous chapter we saw how the association between two measurements, such as the statures of father–son pairs or the heights and weights of a series of individuals, could be calculated in terms of the correlation coefficient. For this procedure to be reasonably satisfactory it was necessary that the two measurements followed a bivariate normal distribution, at least approximately. With heights and weights, for instance, and many other physical measurements, the assumption is probably not too far from the truth. In general, if we select individuals at random and make measurements on them, or alternatively select some randomly chosen unit such as a family, then the pairs of readings certainly have a bivariate distribution of some kind. With luck it will also be approximately bivariate normal. If, however, we specially select individuals on the basis of one measurement and afterwards record the other, the first measurement properly speaking has no distribution: it may have been decided quite arbitrarily.

Thus, the data in Table 14 are quite suitable for the correlation type of analysis: a family is randomly chosen, and the father and one son (also randomly chosen) are measured. Suppose now that we were specially interested in the sons of very tall and very short fathers. We might well consider that Table 14 contained insufficient material about these extremes, and might decide to swell the sample by an extra 100 fathers of about 1.58 m in height and another 100 of about 1.84 m. The observations could still be set out in an array like Table 14, and would show very clearly the association between the statures of fathers and sons. But the bivariate normal distribution would no longer be an even approximately valid assumption, because of the large concentrations of frequencies in the top left-hand and the bottom right-hand corners of the new table, as well as in the tails of the marginal distributions.

Difficulties such as those just described are best dealt with in terms of regression rather than correlation. Indeed, so much experimental research work involves either an obviously non-normal distribution for one set of measurements, or a completely arbitrary series of values (possibly chosen for good extraneous reasons), that it is on the whole the rule for regression analysis to be appropriate and the exception for correlation methods to apply.

Let us consider some data collected by Dr R. E. Moreau on various species of African populations of the bird *Zosterops*. This was a very extensive investigation that was specially concerned with the relation of body measurements to climatic conditions. Many thousands of readings were made and a comprehensive analysis of the correlation or regression type would have been prohibitively laborious. An appropriate simplification or 'reduction of the data' was made possible by the following device. About 80 distinct populations could be distinguished. For each of these the *average* tail, wing and beak lengths were calculated, and the *average* local altitude and maximum and minimum temperatures were also estimated. *These averages were then taken as the basic observations*. We shall for the moment concentrate only on the relation between tail length and altitude. A scatter diagram showing these data in graphical form is presented in Fig. 2.

From Fig. 2 we obtain an immediate impression of a great deal of variation, although there is perhaps a general trend of about

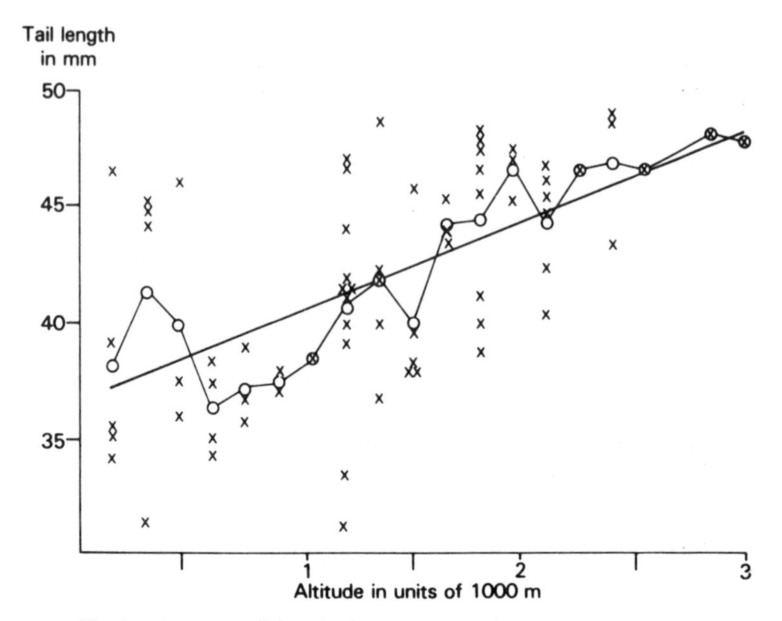

Fig. 2. Average tail lengths for a number of populations of birds plotted against average altitude. Each cross indicates the average tail length (y) at a particular estimated average local altitude (x) for a specific population. Small circles represent the means $\bar{y}_x$ for each value of x. The estimated regression line is also shown.

1 mm increase in tail length for each additional 300 m in altitude. The relative concentration of observations at around 500 m suggests that a bivariate normal distribution is unlikely to be satisfactory. And in any case we are interested not so much in merely establishing an association between tail length and altitude, but in speculating on the effect of changing altitude on tail length. From this point of view the distribution of altitudes is virtually arbitrary. Also, if we had the opportunity to collect further data, it would be much better to fill in the poorly represented altitudes around 800 m and over 2500 m than to have more observations in the centre of the range.

This brings us to the basic characteristics of a simple regression set-up. We assume, for any arbitrary or fixed value of x, that the corresponding values of y that could turn up in repeated samples have a normal distribution about some theoretical mean (say μ_x). A graph showing how these means varied with x would exhibit the underlying relationship between y and x, which is liable to be obscured in practice by large statistical fluctuations. Thus the observed means $\bar{y}_x$, indicated by small circles in Fig. 2, are subject to considerable variation since they are for the most part based on very small numbers of observations. Quite often it is possible to assume that the true y-means for each value of x all lie on a straight line. This is then the *regression line* which exhibits the linear regression of y on x (as distinct from the regression of x on y, which may or may not exist). The slope of the line is the *regression coefficient* of y on x, and measures the average amount by which y increases for a unit increase in x. The coefficient may be positive or negative, or zero if x and y are really unrelated.

Regression analysis therefore turns on discovering estimates of such coefficients and of the additional constant required to fix the position of the regression line. Significance tests are required to see whether the estimated regression coefficient departs from zero (when x and y are independent) more than could be expected by chance. One further assumption is required to carry out such a test: not only must the distribution of y for any given x be normal, but all the distributions for different values of x must have the same variance. This is not such a serious limitation as might be imagined, and the assumption is frequently justified, at least approximately. Inspection of Fig. 2 shows no very striking changes in variance, although if we had more observations at about 2500 m we might find a more restricted distribution than is apparent at, say, 1250 m. The latter may, however, have a greater actual range because of the larger sample available.

Sometimes the regression line is no longer straight but is some kind of curve. With luck we can turn this into the simpler case of linear regression by using a different scale of measurement for one or both of the variables, e.g. taking square roots or logarithms. One must be guided in this rather largely by experience and the observations themselves. Elaborate mathematical procedures are available for fitting 'curvilinear' regressions, or for examining various transformations that might be used. In elementary applications, however, the best plan is to start with a scatter diagram like that in Fig. 2. If it looks as though a straight line is unlikely to be roughly all right, we try a change of scale as suggested above. Perhaps the y-means appear to increase more rapidly with large x. In which case, working with $\sqrt{y}$ or $\log y$ might improve things. If we arrive at a state of affairs where the regression line may well be straight but the variance of y changes with x, we are still in trouble. When unavoidable difficulties like this arise, and they are not as frequent as one might suppose, it is time to seek expert statistical advice.

In this chapter we shall confine our attention to the simplest type of linear regression, i.e. we are assuming that the true y-means lie on a straight line when plotted against x, and that the variance of y for any given x has some constant value, say σ^2. Although in one way these assumptions might seem to be rather restrictive, they are in fact very much less narrow than those entailed by the use of correlation methods. The regression approach is accordingly more generally valid and is widely applicable.

10.2 THE CALCULATION OF REGRESSION COEFFICIENTS

We now turn to the problem of actually fitting a regression line. This entails estimating the constants in the appropriate equation, including the all-important regression coefficient. The question of significance will be dealt with in the following section. Although regression is, as we have seen, logically distinct from correlation, the basic calculations involving sums of squares and products, etc., are precisely the same. This means that there is little fresh to learn from the computational point of view. The basic data are a series of paired values (x, y). (Remember, however, that in the illustration used in the last section each y was really the average tail length for a population of birds and each x was the average estimated altitude for the population.) We first plot these numbers in a scatter diagram like that in Fig. 2 to ensure that the conditions for the validity of a regression analysis are not flouted too outrageously. (If they are we try the devices suggested at the end of the last section.) Next, we perform the calculations described in section 9.2, in particular all those indicated by formulae (35)–(38).

Let us suppose that the equation of the true regression line is given by the expression

$$y = \alpha + \beta x, \tag{43}$$

where, for any given x, y is equal to the corresponding true y-mean μ_x. In formula (43) the symbol β is, in fact, the true regression coefficient of y on x. It is estimated from the sample by b, which is given by either of the two expressions

$$\left.\begin{aligned} b &= \frac{\sum(x-\bar{x})(y-\bar{y})}{\sum(x-\bar{x})^2} \\ &= \frac{c}{{s_x}^2}. \end{aligned}\right\} \tag{44}$$

The constant α is now estimated by

$$a = \bar{y} - b\bar{x}. \tag{45}$$

The fitted regression line is thus

$$y = a + bx, \tag{46}$$

where a and b are given by formulae (45) and (44).

If we decide to look at the data of Table 14 from the standpoint of regression, we find that the regression of son's stature on father's stature is 3.49/7.11 = 0.491. This means that if we measure the heights of the sons in successive groups of fathers, then for each increase of 10 mm in the fathers' heights there is an average increase of about 5 mm in the sons' heights.

We shall now consider the data illustrated in Fig. 2. As the principles of the calculations involved have already been described and illustrated, we shall not give the details, but shall merely quote the basic numerical quantities derived directly from the data:

$$\left.\begin{array}{lll} n = 72 & \bar{x} = 1.338 & \bar{y} = 41.69 \\ \sum(x - \bar{x})^2 & \sum(y - \bar{y})^2 & \sum(x - \bar{x})(y - \bar{y}) \\ \quad = 35.136 & \quad = 1526.30 & \quad = 133.80 \\ s_x{}^2 = 0.49487 & s_y{}^2 = 21.497 & c = 1.8845 \end{array}\right\}$$

The constants α and β are therefore estimated by

$$b = \frac{133.80}{35.136} = 3.8081,$$

and

$$a = 36.595,$$

so that the regression line is

$$y = 36.60 + 3.808x.$$

This line is easily transferred to the graph by drawing a line through any pair of points on it, preferably chosen as far apart as possible. Thus if $x = 1$ we find, by substituting in the equation, that $y = 40.41$; and similarly if $x = 3$ we have $y = 48.02$. These two points determine the regression line, which is shown in Fig. 2.

We have not yet considered estimating the variation about the regression line. The analysis assumes that, for any given x the sample values of y have a normal distribution with mean given by formula (43). Moreover, all distributions have the same variance, σ^2. This is estimated by s^2, where

$$s^2 = \frac{1}{n-2}\left\{\sum(y - \bar{y})^2 - \frac{\left[\sum(x - \bar{x})(y - \bar{y})\right]^2}{\sum(x - \bar{x})^2}\right\}. \tag{47}$$

Points to note in formula (47) are the divisor $n - 2$ outside the braces, and the *square* of the sum of products about means appearing inside. These calculations involve only the basic sums of squares and products with which we are already familiar. It is clear from the form of formula (47) that, for reasonably large n, s^2 must be less than $s_y{}^2$, which corresponds to the use of only the first term within the braces. This is what we should expect, since fitting the regression line removes some of the total variation of y. The results shown above for the data on which Fig. 2 is based give

$$s^2 = \frac{1}{70}\left\{1526.30 - \frac{(133.80)^2}{35.136}\right\}$$
$$= 14.53.$$

Thus, the distribution of average tail length for any chosen value of altitude has the mean given by the appropriate point on the regression line and standard deviation 3.81. Suppose we take populations at 2000 m. Then with $x = 2$ we have $y = 44.22$. We can then describe the variation in average tail length between populations by a normal distribution with mean 44.22 and standard deviation 3.81. These are the sort of numerical results required for discussion of questions like the adaptation of a species to the aerodynamical conditions entailed by the environment.

It should be emphasised that, as with the calculation of correlation coefficients in the previous chapter, a good pocket calculator should require only the entering of the sequence of (x, y) pairs. After that most of the above quantities (means, standard deviations, regression estimates, etc.) should be automatically available. But you might have to calculate s^2 yourself from formula (47) using the results provided. In any case be *extremely* careful that you have obtained the regression of y on x and not the regression of x on y.

10.3 STANDARD ERRORS AND SIGNIFICANCE TESTS

In the data exhibited in Fig. 2 it is fairly clear, in spite of the considerable amount of natural variation present, that there *is* a definite relationship between tail length and altitude. We should be very surprised if told that this apparent association was not statistically significant. However, cases often arise where the issue is in doubt, and a proper significance test is then required.

In order to test for the existence of some degree of association, we need to know whether the observed regression coefficient b is significantly different from the hypothetical value zero. On the assumption made, b is normally distributed with mean β and variance $\sigma^2/\sum(x - \bar{x})^2$, but we usually do not know σ and have to estimate it by s. If n is fairly large, say greater than 30, then we can write the estimate with an attached standard error as

$$b \pm \frac{s}{\sqrt{\left[\sum(x - \bar{x})^2\right]}}. \tag{48}$$

The ordinary test for normally distributed estimates then applies: e.g. if b is more than 1.96 standard errors away from zero, then it is significant at the 5 per cent level. We can also test departures from any other hypothetical value of the regression coefficient that is different from zero in precisely the same kind of way. If a well-established coefficient has value β or if this is the figure predicted by some theory, then we examine the difference $b - \beta$ in terms of the same standard error as that shown in formula (48).

In the illustration of the last section involving the relationships between tail length and altitude, we find that $s/\sqrt{[\sum(x - \bar{x})^2]} = 0.643$. We can therefore write the regression coefficient as

$$b = 3.808 \pm 0.643.$$

The number of observations in the sample is quite large, and so we can use the simplest test for a normal variable. The estimate is, in fact, between five and six standard deviations from zero, and the regression is therefore, as expected, highly significant.

When n is smaller than about 30, we have a situation which is exactly analogous to those discussed in Chapter 6 on the use of t-tests. To test whether an observed b is significantly different from a hypothetical value β, we use

$$t = \frac{b - \beta}{s\Big/\sqrt{\left[\sum(x - \bar{x})^2\right]}}, \tag{49}$$

where the number of degrees of freedom is $n - 2$.

We thus see that the regression coefficient can be tested in the same way as the means of normal samples already discussed in Chapters 5 and 6. Confidence limits can also be attached using the methods described earlier in sections 3.2 and 6.4. In the example above we can use 'normal' limits because the sample is large. Thus, we have $3.808 \pm (1.96 \times 0.643)$, or 2.548 to 5.068 for a 95 per cent confidence interval.

When samples are small we must resort to the use of the t-distribution.

Comparison of two regression coefficients

Another fairly simple test that is often required is the comparison of two regression coefficients. Thus, the African species of *Zosterops* has been shown to have a regression coefficient of 3.808 for tail length on altitude. We might also have collected data on the species prevalent in some other large geographical area and wish to know whether the dependence of tail length on altitude was the same. When numbers are large this involves no more than comparing two normally distributed estimates, both of whose standard errors are known. Expressions such as formula (48) are calculated for each sample separately. The variance of the difference between the two estimates of regression coefficients, i.e. $b_1 - b_2$, is then the sum of the individual variances. We therefore calculate

$$d = \frac{b_1 - b_2}{\sqrt{\left[\dfrac{s_1^2}{\sum_1(x - \bar{x}_1)^2} + \dfrac{s_2^2}{\sum_2(x - \bar{x}_2)^2}\right]}}, \qquad (50)$$

where the suffixes 1 and 2 refer to the two samples, and d is normally distributed with zero mean and unit standard deviation. Subsequent procedure is then similar to that explained in section 5.2 for comparing the *means* of two large samples.

When numbers are small the required test is a simple modification of the t-test for comparing the means of two samples, as described in section 6.3. The analogue of formula (21) is

$$\left.\begin{aligned} t &= \frac{b_1 - b_2}{s\sqrt{\left[\dfrac{1}{\sum_1(x-\bar{x}_1)^2} + \dfrac{1}{\sum_2(x-\bar{x}_2)^2}\right]}}, \\ \text{where} \\ s^2 &= \frac{(n_1-2){s_1}^2 + (n_2-2){s_2}^2}{n_1+n_2-4}, \end{aligned}\right\} \qquad (51)$$

the number of degrees of freedom being $n_1 + n_2 - 4$. The pooling of the sums of squares in this test is really only legitimate if the variation about the regression line is the same for each sample. We can test this by examining the variance-ratio, $F = {s_1}^2/{s_2}^2$, as mentioned in section 6.5. If a significant result were obtained, we could not validly employ the t-test as above, and should be obliged instead to resort to an analogue of the modification described in section 6.5. We should then calculate d as in formula (50) above, but should treat it as a 'Student's' t with f degrees of freedom given by

$$\left.\begin{aligned} f &= \frac{1}{\dfrac{u^2}{n_1-2} + \dfrac{(1-u)^2}{n_2-2}}, \\ \text{where} \\ u &= \frac{{s_1}^2/\sum_1(x-\bar{x}_1)^2}{{s_1}^2/\sum_1(x-\bar{x}_1)^2 + {s_2}^2/\sum_2(x-\bar{x}_2)^2}. \end{aligned}\right\} \qquad (52)$$

Confidence limits for the difference $\beta_1 - \beta_2$ are obtained as usual with t-variables.

11

Simple experimental design and the analysis of variance

11.1 INTRODUCTION

In most of the previous chapters the main object has been to analyse scientific data so as to estimate important constants or to carry out tests of significance. The data may have arisen more or less by chance, and must be examined for signs of bias or heterogeneity. Careful sifting of the material may help to eliminate some errors of this sort, but we must guard against introducing fresh ones. In examples like that of section 5.2, where the body-weights of male and female birds were compared, we have so far assumed that the observations have been collected in such a way as to avoid unwanted bias. Sometimes, however, it is possible to exercise stricter control over the data so that these are obtained under carefully specified conditions. This leads naturally to the whole question of designing the general pattern of an experiment, partly to eliminate various disturbing effects that might creep in and partly to ensure that maximum precision is achieved for the amount of effort expended.

Some experiments, such as those whose main purpose is to determine a specific number, for example, a recombination fraction in genetics or the generation-time of a dividing bacterium, are clearly 'absolute' in character. Others, which are concerned with comparing the effects of different factors or treatments are of a 'comparative' nature. What is usually known as *experimental design* deals with methods of constructing and analysing these comparative experiments. An enormous range of experimental patterns has been developed in recent years, and designs are available to suit a wide variety of circumstances. Fortunately, the majority of everyday needs can be met sufficiently well by a fairly small number of designs. This chapter and the next deal with only the most elementary arrangements that arise in practice. Apart from introducing the reader to the underlying principles of the subject as a basis for further reading, the discussion will also provide a number of basic lay-outs, which are capable of facilitating the analysis of a great many frequently occurring situations.

The whole subject of experimental design may be considered under two main headings. The first is the problem of the actual pattern of the experiment. A correct choice of design depends partly on a knowledge of the experimental material and partly on the kind of questions one wishes to ask. Secondly, there is the analysis of the data. As a rule, the results of the kind of experiment of which we are thinking can be neatly summarised in what is called an *analysis of variance table*, together with a table of treatment means and their standard errors. Such tables, moreover, are based on nothing more complicated than suitably chosen sums of squares, thus involving statistical calculations with which we are already familiar.

The design and analysis of an experiment obviously influence one another very strongly. For a given design, there is usually only one really satisfactory way of analysing the data and presenting the results. If the design is badly chosen, the analysis may be unduly laborious, and it may even turn out that certain important questions cannot be answered at all. We therefore want to aim at designs that are simple in character and are easily and unambiguously analysable. For these reasons it is extremely important that, where possible, alternative designs be carefully considered before embarking on the actual procedure of experimentation. If there is any doubt as to the proper course of action or if difficulties are likely to arise, it is highly desirable to seek expert statistical advice *before the experiment has been started*. Few things are more galling than to carry out an elaborate and, in detail, technically sound experiment, only to find out afterwards that some mistake in the pattern of the design vitiates the results obtained. Sometimes shortcomings in the design can be overcome by persuading a biostatistician to embark on a more subtle form of analysis. But all too often the comparisons sought are inextricably mixed with other factors, and nothing short of a new experiment is of any avail.

Since the theory and practice of experimental design have been developed largely in the context of agriculture, we shall use illustrations of a type primarily belonging to this subject. Applications may, however, be made over a very much larger area of research. Some of the special problems arising in other fields are mentioned explicitly in section 11.5.

11.2 COMPLETELY RANDOMISED DESIGNS

We have already seen some examples of very simple experimental designs in sections 6.2 and 6.3, where t-tests were used to compare *two* treatments or *two* samples of observations. The first extension of these methods is to see what happens where several groups are involved. Suppose, for example, we wish to compare the yields of k varieties of wheat. One way of doing this would be to allot to each variety a certain number of small plots of standard size and to take the yield of a single plot as the basic variable x. The performance of the ith variety would be indicated by the mean yield, say $\bar{x}_i$ based on n_i plots. Comparisons between the varieties would therefore depend on comparisons between the k means. In the first instance we should merely want to know whether any significant differences could be detected between the means. If differences were present, the individual estimates of the average yields with standard errors would be required.

Because quite large variations in soil fertility are often present, the very simple set-up suggested above might be rather inefficient in practice. Nevertheless, it would be perfectly valid provided that the available plots of land were assigned randomly and without bias to the several varieties. In the next section we shall see how to eliminate some of the variation between areas of soil by the use of specially chosen blocks of plots. For the moment we shall suppose that this was not necessary or was for some reason impossible. The basic data obtained from this type of design are shown in Table 15.

The suffix notation is unavoidable if we want to exhibit the form of the data quite generally, as in Table 15. However, most of the calculations involved can be simply explained. First, we notice that no special assumption is made about the number of plots allotted to each variety: they may or may not all be equal. The ith variety has n_i observed yields totalling T_i, so that the variety-mean is $\bar{x}_i = T_i/n_i$. There are $N = \sum n_i$ observations in all with a grand total of all yields $G = \sum T_i$. The mean of all the observations is then $\bar{x} = G/N$.

When comparing the means of *two* samples in section 6.3, we assumed that the individual variances were the same even though the means might differ. We shall make a similar assumption here, namely that the true variance for each variety is the single quantity σ^2. In order to test for difference between the observed variety-means $\bar{x}_i$, we adopt the obvious null hypothesis that all means are equal. Now we can make two estimates of σ^2. One of these, s^2, is based, as in section 6.3, on the pooled sums of squares for each variety abouts its own mean. The other is obtained from the set of means $\bar{x}_i$. If the variance-ratio test shows that the latter estimate is significantly greater than the former (but not vice versa), we conclude that the variation between the varietal means is more that we should expect by chance.

The quickest and most satisfactory way to calculate the required quantities is to work as follows. We first find the total sum of squares about the general mean $\bar{x}$. This is

$$\left.\begin{aligned} \sum(x_{ij} - \bar{x})^2 &= \sum x_{ij}^2 - C, \\ \text{where} \qquad & \\ C &= G^2/N. \end{aligned}\right\} \qquad (53)$$

Summation in the first line of formula (53) is over all N observa-

Table 15. *General table of data from a completely randomised design*

Variety	Observed plot yields	No. of obs.	Total	Mean
1	$x_{11}, x_{12}, \ldots, x_{1n_1}$	n_1	T_1	$\bar{x}_1$
2	$x_{21}, x_{22}, \ldots, x_{2n_2}$	n_2	T_2	$\bar{x}_2$
.	.	.	.	.
.	.	.	.	.
.	.	.	.	.
i	$x_{i1}, x_{i2}, \ldots, x_{in_i}$	n_i	T_i	$\bar{x}_i$
.	.	.	.	.
.	.	.	.	.
.	.	.	.	.
k	$x_{k1}, x_{k2}, \ldots, x_{kn_k}$	n_k	T_k	$\bar{x}_k$
Total		$N = \sum n_i$	$G = \sum T_i$	$\bar{x} = G/N$

tions, and we are merely using the equivalent of formula (7) in Chapter 2. The correction factor C is worth recording separately as we shall need it again.

Secondly, we compute

$$\sum T_i^2/n_i - C, \tag{54}$$

where summation is of course over the k varieties. Notice that the divisor of each squared total in formula (54) is precisely the number of observations making up that total. We can now start filling in the second column of the analysis of variance table shown in Table 16.

This table follows a certain standard pattern which should always be used. What we are doing is to analyse the total sum of squares, $\sum x_{ij}^2 - C$, into two components. One is the part corresponding to variation between the variety-means, $\sum T_i^2/n_i - C$, and the other is the so-called 'residual' or 'error' sum of squares. The latter is most easily found by subtracting the variety sum of squares from the total, and it arises in practice from variations in the experimental material, 'errors' in technique and so on. In this simple analysis of variance the residual sum of squares can easily be found directly by the alternative procedure of calculating the

Table 16. *Analysis of variance for Table 15*

Source of variation	Sum of squares	Degrees of freedom	Mean squares
Between varieties	$\sum T_i^2/n_i - C$	$k-1$	M
Residual	By subtraction	$N-k$	s^2
Total	$\sum x_{ij}^2 - C$	$N-1$	–

sum of squares for each variety separately, and then adding, i.e.

$$\sum_{i=1}^{k}\left\{\sum_{j=1}^{n_i} x_{ij}^2 - T_i^2/n_i\right\}. \tag{55}$$

This makes quite a useful check, though for more complicated types of analysis the labour of the direct calculation is usually prohibitive.

We now turn to the degrees of freedom, which are an essential part of the analysis of variance technique. The total sum of squares about the over-all mean has $N-1$ degrees of freedom, corresponding to the divisor normally used in calculating an estimate of variance. There are $k-1$ degrees of freedom for the sum of squares relevant to fluctuations between k varieties. Subtraction then gives the degrees of freedom for the residual, i.e. $(N-1)-(k-1)=N-k$. These degrees of freedom are to be used as divisors to obtain the 'mean squares' given in the last column of the table. Thus, the estimate of the σ^2 obtained from the residual is $s^2 = (\text{residual sum of squares})/(N-k)$; and M, the estimate derived from the differences between variety means, is $M = (\sum T_i^2/n_i - C)/(k-1)$.

The table is now complete. The quantity M is a measure of the 'spread' of the variety means, and if it is sufficiently greater than s^2, we regard the result as significant: the null hypothesis must be rejected, and at least some real differences between varieties admitted. (No particular importance attaches to values of M smaller than s^2, though such results would be suspicious if they occurred frequently.) The appropriate statistical test is the variance-ratio test already mentioned in section 6.5. However, in the application considered in that section a two-tailed test was required and we had to divide the larger of the two variances by the smaller. On the other hand, the present context demands a one-tailed test. We calculate

$$F = \frac{M}{s^2}, \tag{56}$$

where the number of degrees of freedom for the numerator is $f_1 = k - 1$, and for the denominator is $f_2 = N - k$. We then refer to tables of the F-distribution, using the significance levels as they stand and not doubled as in section 6.5. Because of the widespread use of this test in the analysis of experimental designs, both 5 per cent and 1 per cent points of the variance ratio are given at the end of this book in Appendix 5. Other convenient references are in Fisher & Yates' *Statistical Tables* (where F is called e^{2z}) and in *Biometrika Tables for Statisticians*.

A numerical illustration should make clear just what is involved in these calculations. Table 17 gives the yields of grain in kilograms of plots of 100 m^2 for four varieties of wheat. It looks on the face of it as though variety No. 2 is giving a rather higher yield than the others, and that No. 3 is a somewhat poorer type. A fair amount of variation is present, and so we must decide whether this is sufficient to account for the differences observed between varieties.

The calculations required are as follows:
Total number of observations, $N = 19$.
Grand total, $G = \sum x_{ij} = 697.7$. Correction factor, $C = G^2/N$

$$= 25\,620.28.$$

Table 17. *Yields in kg per plot of 100 m^2 for four varieties of wheat*

Variety	Plot yields (x_{ij})	No. of obs (n_i)	Total (T_i)	Mean ($\bar{x}_i$)
1	34.3, 35.7, 37.8, 36.9, 33.2	5	177.9	35.58
2	40.4,38.0,38.3,40.6,38.8,43.5	6	239.6	39.93
3	30.5, 32.0, 33.4, 36.2	4	132.1	33.02
4	33.1, 38.4, 35.7, 40.9	4	148.1	37.02
Total		19	697.7	36.72

Total sum of squares about the mean,

$$\sum x_{ij}^2 - C = (34.3)^2 + (35.7)^2 + \ldots + (40.9)^2 - 25\,620.28$$

$$= 210.21.$$

Sum of squares between varieties,

$$\sum T_i^2/n_i - C = \frac{(177.9)^2}{5} + \frac{(239.6)^2}{6} + \ldots + \frac{(148.1)^2}{4} - 25\,620.28$$

$$= 123.44.$$

We can start filling in the analysis of variance table shown in Table 18, completing the residual sum of squares by subtraction.

It is immediately obvious from Table 18 that the varietal mean square is higher than the residual. Using formula (56) we have $F = M/s^2 = 7.11$. The 5 per cent point corresponding to 3 and 15 degrees of freedom, *in that order*, is 3.29, while the 1 per cent point is 5.42, so there are certainly significant differences between the means at the latter level.

Having established significant differences between the varieties, we begin to be more interested in estimating the actual means with their standard errors. In general, we can write the mean yield of the ith value as

$$\bar{x}_i \pm s/\sqrt{n_i}. \tag{57}$$

If $N - k$ is fairly large we can treat $s/\sqrt{n_i}$ as an accurate estimate of the standard error of $\bar{x}_i$, and obtain confidence limits on the basis of the normal distribution as in section 3.2. If, however, $N - k$ is small then we must use the t-distribution as described in section 6.4.

In the above example we have $s = \sqrt{5.79} = 2.41$. The four means can therefore be written with their standard errors as in Table 19.

Although the general significance test points to real differences between the varieties, it does not automatically follow that every variety is different from every other. Thus 1, and 3 are unlikely to have a big real difference, although 2 and 3 are substantially

Table 18. *Analysis of variance for Table 17*

Source of variation	Sum of squares	Degrees of freedom	Mean square	Variance ratio
Varieties	123.44	3	41.15	7.11
Residual	86.77	15	5.79	–
Total	210.21	18	–	–

Table 19. *Variety means with standard errors for Table 17*

Variety	Mean and standard error
1	35.58 ± 1.08
2	39.93 ± 0.98
3	33.02 ± 1.20
4	37.02 ± 1.20

dissimilar. We can compare any specific pair of varieties, say i and j, by using the mean difference with its standard error

$$(\bar{x}_i - \bar{x}_j) \pm s\sqrt{\left(\frac{1}{n_i} + \frac{1}{n_j}\right)}, \qquad (58)$$

and referring to the normal or t-distribution (with $N - k$ degrees of freedom) as appropriate.

Care is needed in testing specific comparisons unless these are stipulated *before* the experiment is performed. It should be obvious that if there is a large number of varieties, then even when the results of the analysis of variance test are not significant, the largest and smallest means may be widely divergent. We are then liable to accept such differences as real when they are due only to sampling variation. It is, however, legitimate to arrange all means in order of magnitude, and regard two consecutive means as significantly different if a significant result is obtained from the test indicated by formula (58).

There are special methods of coping with this situation in detail, but they are beyond the scope of the present treatment.

Advantages and disadvantages of the completely randomised design

There are a number of special characteristics possessed by completely randomised designs to which it is worth drawing attention. These are summarised below:

(i) The designs are very flexible and can be used for any number of treatments, and may have any numbers (not necessarily all the same) of observations in each treatment group.

(ii) The statistical analysis is comparatively easy and straightforward. It is, moreover, unaffected if some or all of the observations for any treatment are lost or missing for some purely random accidental reason, i.e. if the accident is not more likely to happen to one treatment rather than another. We merely carry out the standard analysis on the observations that are available.

(iii) However, in certain circumstances the design suffers from the disadvantage of being inherently less informative than other more sophisticated lay-outs. If there are large differences between blocks, due say to fluctuations in fertility, the whole of this variation is included in the residual variance, making the usual significance tests less sensitive. It is then better to use the randomised block design described in the next section. With entirely homogeneous material, on the other hand, the completely randomised lay-out is the most accurate.

11.3 RANDOMISED BLOCK DESIGNS

Although the completely randomised arrangement of the last section is the simplest possible set-up and has several advantages, it is liable to be rather inefficient if there are big differences in, for example, soil fertility. When this occurs the basic variation represented by σ^2 will contain amongst other things a large element due to the fluctuations in fertility. As a result, small but important differences between variances may be swamped by the general variability unless a very large experiment is planned.

It will be recollected that in section 6.2, where we were comparing the effects of two drugs, large differences between patients could be eliminated by the method of 'paired comparisons'. We first calculated the relative advantage of one drug over the other for each patient separately, then based the analysis on this set of relative advantages. A somewhat similar device can be used where there are several treatments to be tested and where the experimental material or environmental conditions show wide variation.

Suppose we suspect that fluctuations in soil fertility are going to be an important source of variability. We want a design that will separate out this factor, and so enable us to test the treatment differences, in which we are primarily interested, with greater precision. The simplest method is the arrangement called *randomised blocks*. It is well known that if a sufficiently small area of land is used, then fluctuations in fertility can be substantially reduced. If the experiment we have in mind involves a large number of plots, then it is obvious that we cannot expect to accomodate them all in a single homogeneous area. But if we divide up the land into several distinct blocks, it may be possible to arrange that each of these blocks is fairly uniform although there may be large differences between blocks. We then divide up each block into as many plots as there are treatments to be tested. The treatments are allocated at random to the plots in any given block.

In the previous section we envisaged four varieties of wheat: these are the 'treatments'. With a completely randomised design, we could have any number of plots for each variety, but in the randomised block lay-out the numbers must all be the same and equal to the number of blocks. Suppose that in a generalised notation we have t treatments and b blocks; the plot for treatment i in block j has yield x_{ij}; the total yield for the ith treatment is T_i; the total yield for the jth block is B_j; and the grand total of all bt yields is G. Then the data from the experiment can be set out as in Table 20. This should be compared with Table 15. In that table we made special use of the *variety total*, but in the present case we need the *block totals* as well. A rather similar analysis to the previous one can now be carried out.

This time the general correction factor is $C = G^2/bt$. We calculate the total sum of squares about the general mean, $\bar{x}$,

Table 20. *General table of data for a randomised block design*

	Block								
Treatment	1	2	...	j	...	b	Treatment total	Treatment mean
1	x_{11}	x_{12}	...	x_{1j}	...	x_{1b}	T_1	$\bar{x}_1$
2	x_{21}	x_{22}	...	x_{2j}	...	x_{2b}	T_2	$\bar{x}_2$
·	·	·		·		·	·	·
·	·	·		·		·	·	·
·	·	·		·		·	·	·
i	x_{i1}	x_{i2}	...	x_{ij}	...	x_{ib}	T_i	$\bar{x}_i$
·	·	·		·		·	·	·
·	·	·		·		·	·	·
·	·	·		·		·	·	·
t	x_{t1}	x_{t2}	...	x_{tj}	...	x_{tb}	T_t	$\bar{x}_t$
Block total	B_1	B_2	...	B_j	...	B_b	G	$\bar{x} = G/bt$

exactly as before. The sum of squares for treatments is similar to that given in formula (54) with every n_i equal to b, i.e.

$$\left.\begin{aligned} &\sum_i T_i^2/b - C, \\ \text{where} \quad &C = G^2/bt. \end{aligned}\right\} \quad (59)$$

The analogous quantity for blocks is

$$\sum_j B_j^2/t - C. \quad (60)$$

As usual in this type of analysis, the divisors of the squared totals in formula (59) and (60) are exactly equal to the numbers of observations making up these totals. We can thus build up an analysis of variance table, which is analogous to Table 16 but which contains an additional row for block effects. The residual sum of squares is found, as before, by subtracting the other items from the total. And the mean squares are calculated by dividing each sum of squares by the corresponding number of degrees of freedom. Table 21 shows the complete set-up.

Table 21. *Analysis of variance for Table 20*

Source of variation	Sum of squares	Degrees of freedom	Mean square
Treatment	$\sum_i T_i^2/b - C$	$t-1$	M_T
Block	$\sum_j B_j^2/t - C$	$b-1$	M_B
Residual	By subtraction	$(t-1)(b-1)$	s^2
Total	$\sum x_{ij}^2 - C$	$bt-1$	–

We can now test the significance of the block and treatment mean squares by the variance-ratios M_B/s^2 and M_T/s^2. The great advantage of the design is that the block and treatment effects have been completely disentangled in the sense that we can make these two tests quite independently of one another. The ratio M_B/s^2 is very likely to be significant if we are using the design to cope with the kind of difficulty already mentioned, such as large fluctuations in soil fertility. This test tells us, moreover, whether the classification into blocks was worth while, which might be of great practical importance if the division into blocks were very difficult or expensive. Nevertheless, no matter what the result of this test, we can still validly test the treatments; and, if these are significant, go on to compile a table of treatment means, $\bar{x}_i$, each with an attached standard error of $s/\sqrt{b}$, i.e. we can write

$$\bar{x}_i \pm s/\sqrt{b}. \tag{61}$$

The corresponding expression for the difference between any two treatments, say the ith and jth, is

$$(\bar{x}_i - \bar{x}_j) \pm s\sqrt{\frac{2}{b}}. \tag{62}$$

As an illustration, we use the numerical data of Table 22, where we have 5 blocks, each of 4 plots, and 4 varieties to test.

The basic calculations are:

Total number of observations, $bt = 20$.

Grand total, $G = \sum x_{ij} = 722.7$. Correction factor, $C = G^2/bt$

$= 26\,114.764.$

Table 22. *Yields in kilograms per plot of 100 m^2 for four varieties of wheat*

Variety	Block 1	2	3	4	5	Variety total	Variety mean
1	33.4	34.1	35.4	36.7	37.7	177.3	35.46 ± 0.32
2	38.2	38.0	40.6	40.3	43.7	200.8	40.16 ± 0.32
3	31.8	30.5	33.4	32.6	36.2	164.5	32.90 ± 0.32
4	33.2	35.2	35.7	37.1	38.9	180.1	36.02 ± 0.32
Block total	136.6	137.8	145.1	146.7	156.5	722.7	36.13

Total sum of squares about the mean,

$$\sum x_{ij}^2 - C = (33.4)^2 + (34.1)^2 + \ldots + (38.9)^2 - 26\,114.764$$

$$= 205.965.$$

Sum of squares for treatments,

$$\sum T_i^2/b - C = \tfrac{1}{5}\{(177.3)^2 + \ldots + (180.1)^2\} - 26\,114.764$$

$$= 135.673.$$

Sum of squares for blocks,

$$\sum B_j^2/t - C = \tfrac{1}{4}\{(136.6)^2 + \ldots + (156.5)^2\} - 26\,114.764$$

$$= 64.123.$$

The resultant analysis of variance table can now be filled in and completed as shown in Table 23.

It is at once evident that the differences between varieties are highly significant, since the variation ratio F is 88.0 with 3 and 12 degrees of freedom. The block component is also highly significant. A substantial amount of variation has thus been removed, although in this experiment the simpler completely randomised design would still have revealed significant differences between varieties.

Table 23. *Analysis of variance for Table 22*

Source of variation	Sum of squares	Degrees of freedom	Mean square	Variance ratio
Variety	135.673	3	45.22	88.0
Block	64.123	4	16.03	31.2
Residual	6.169	12	0.514	—
Total	205.965	19	—	—

The actual variety means are shown in Table 22. Each of these has standard error $\sqrt{(0.514/5)} = 0.321$, and the standard error of the difference between any two treatments is $(0.321)\sqrt{2} = 0.454$.

Missing-plot technique

If only *one* yield is missing, say that for x_{ij}, we make an estimate given by

$$x'_{ij} = \frac{tT'_i + bB'_j - G'}{(t-1)(b-1)}, \tag{63}$$

where T'_i, B'_j and G' are the treatment, block and grand totals for the observations actually available. The analysis is then carried out as before as though x'_{ij} were a real observation, with the proviso that the total sum of squares and the residual sum of squares each lose one degree of freedom.

If more than one yield is missing, a special modification of the above method is required.

Advantages and disadvantages of randomised blocks

Randomised block designs are probably the most widely used and have several points in their favour:

(i) With heterogeneous material the residual variance can be reduced by choosing blocks of plots such that the plots within any block are fairly similar, although big differences may occur between blocks.

(ii) There is no restriction on the numbers of blocks or treatments, but in each block there must be the same number of plots, one to each treatment.

(iii) If some yields are accidentally lost, the analysis is again without due complications, although special modifications are required.

11.4 TESTING THE HOMOGENEITY OF VARIANCES

In applying the experimental designs described above, we have assumed that the variance of the uncontrolled fluctuation was everywhere the same. This means that in the completely randomised lay-out of section 11.2, for example, we assume that the variances of each treatment groups taken separately are the same. It might easily happen that the differences between treatments were of a kind that involved larger or smaller fluctuations, quite apart from any differences in the general level of yield. Again, in the randomised block design we might manage to choose several homogeneous blocks, but plots from a block with a relatively low fertility might show less natural variation than plots from a more fertile block.

When such differences in variance exist, we cannot legitimately employ the foregoing omnibus analyses in which the residual variance is estimated jointly from all parts of the lay-out. We can of course test the means of any pair of treatments by using the modifed procedure mentioned in section 6.5. However, the simplified analysis of variance technique can sometimes be retained by using a different scale of measurement. That is, we work with some suitable mathematical function of the observed measurements, such as square roots or logarithms. This will often do the trick, especially if the blocks with the highest means also have the largest individual variances. Similar ideas have already been mentioned in section 10.1 in connection with regression analysis. Such applications are probably beyond the scope of elementary analysis, and statistical advice should be sought if they seem to be necessary.

If inspection of the data suggests that some heterogeneity of variance may be present, then we should probably feel obliged to try to test the suggestion. A suitable method is Bartlett's test for the homogeneity of variances. Suppose there are k groups as in section 11.2. We calculate the variance for each group separately, i.e. in the ith group we have

$$s_i^2 = \frac{1}{n_i - 1}\left\{\sum_{j=1}^{n_i} x_{ij}^2 - T_i^2/n_i\right\}. \qquad (64)$$

We also need the estimate of variance s^2 as calculated for Table 17. If we are not thinking specifically in terms of a completely randomised design, but just want to compare k different estimates of variance based on independent samples, we may prefer to use the alternative formula

$$s^2 = \frac{\sum f_i s_i^2}{f},$$

where

$$f_i = n_i - 1, \text{ and } f = \sum f_i. \qquad (65)$$

Bartlett's test now consists in calculating

$$\frac{1}{C}\left\{f \log_{10} s^2 - \sum f_i \log_{10} s_i^2\right\},$$

where

$$C = 0.4343\left[1 + \frac{1}{3(k-1)}\left\{\sum \frac{1}{f_i} - \frac{1}{f}\right\}\right]. \qquad (66)$$

The quantity in the first line of formula (66) is distributed approximately like χ^2 with $k - 1$ degrees of freedom, and this is tested for significance in the usual way.

Take, for instance, the data of Table 17. We have the individual variances and degrees of freedom shown in Table 24. We also have $f = 15$, and $s^2 = 5.79$. Substitution in formula (66) gives $C = 0.4850$, so that

$$\chi^2 = \frac{15 \log_{10} 5.79 - 4 \log_{10} 3.50 - \ldots - 3 \log_{10} 11.36}{0.4850}$$

$$= 1.370,$$

with 3 degrees of freedom.

Table 24. *Individual variety variances calculated from Table 17*

i	s_i^2	f_i
1	3.50	4
2	4.22	5
3	5.88	3
4	11.36	3

The value of χ^2 just obtained is not significant, and we conclude that the variances of the four groups are homogeneous so far as we can tell from such small samples. There is no reason to suppose that the big difference in estimated variance between the first and last groups is due to anything other than sampling variation. Since the variances pass the homogeneity test, we feel more confidence in the validity of the analysis of variance used in section 11.2. Strictly speaking, we always ought to carry out such a test of variance before embarking on the analysis of variance. But it is often omitted, especially for designs that are more complicated than the completely randomised arrangement, partly on grounds of the time involved, partly because of the greater complexity of Bartlett's test, and partly because moderate amounts of variance heterogeneity are unlikely to have a major influence on the analysis of variance. A warning should, moreover, be given that Bartlett's test is rather sensitive to departures from the assumption of an underlying normal distribution, and may, if care is not taken, give rise to misleading results.

11.5 GENERAL REMARKS

Although the simple experimental designs described in this chapter have been discussed primarily in relation to agricultural research, which is the field in which they were originally developed, applications can be made to most branches of biology. In agriculture the blocks consist of plots of land; but in other sciences a block of homogeneous material might be a rack of test-tubes in an incubator, a number of experiments performed on a particular day by a particular technician, several animals from a single litter, a pair of twins, a single individual on whom several different tests are made, or in fact any group of experimental units expected to possess some degree of homogeneity. We might consider, for example, the paired-comparison test of section 5.1, where each individual was subjected to two drugs in turn, and in which we were virtually using a randomised block design with two 'plots' in each block.

Another point worth making is that we can analyse the effect of *two* factors simultaneously by regarding one of them, e.g. varieties of wheat, as the set of 'treatments' in a randomised block design, and interpreting the other, e.g. alternative types of fertiliser, as the 'blocks'. The significance of differences between varieties is examined by testing the 'treatment' mean square against the residual, and differences between fertilsers by testing the 'blocks' mean square in the same way. We can obtain fertiliser means with standard errors in a manner analogous to that used for varieties. The advantage of employing blocks to eliminate large variations in the experimental material has of course been lost. But with fairly homogeneous units there would be no serious disadvantage.

A slightly more advanced design is available in which there are several observations (preferably constant) for each variety–fertiliser combination. This allows us to find out whether the two factors operate independently of one another or whether they interact to some extent. The problem of examining several factors simultaneously and investigating the question of interactions in particular will be considered in more detail in the next chapter on factorial experiments.

Finally, it must be said that it is most important, from the point of view of understanding and interpretation, to appreciate the details of the computations involved. For small designs with not too much data the formulae can be used as shown, employing a pocket calculator to carry out the arithmetic. But for large and more ambitious designs with big data sets it is highly desirable to employ a fully computerised approach using specialised statistical software.

12

Introduction to factorial experiments

12.1 THE FACTORIAL PRINCIPLE

In the experimental lay-outs, such as the completely randomised and randomised block designs of Chapter 11, the main object is to compare and estimate the effect of a single set of 'treatments' such as different varieties of wheat. The chief purpose of introducing blocks is to make allowance for unwanted but unavoidable heterogeneity. However, as pointed out in section 11.5, we can in some cases use a randomised-block type of analysis for an experiment in which the blocks merely represent the levels of a second factor, such as different types of fertiliser. This raises the whole question of how many factors can be incorporated in a single experimental design, and whether this is or is not a desirable practice.

It is customary in the classical type of scientific experimentation to advocate the investigation of any problem by holding most of the variable factors constant and allowing only one or two to vary in each experiment. If one is largely occupied with fundamental research, where the point is to formulate general laws and test crucial predictions, this procedure of isolating one or two factors at a time has much to recommend it.

If, on the other hand, one is dealing with work that is of a more general nature, such as the prosecution of a plant-breeding programme, then one is essentially concerned to know what happens with a range of combinations of factors. A series of experiments in which only one factor is varied at a time would be both lengthy and costly, and might still be unsatisfactory because of systematic changes in the general background conditions. An alternative approach is to try to investigate variations in several factors simultaneously. This leads us to the idea of a *factorial experiment* in which the set of experimental units, e.g. cultivated plots, animals, Petri dishes, etc., is made large enough to include all possible combinations of levels of the different factors. Thus, if we had three varieties of wheat and three different levels of a fertiliser, there would be nine combinations in all. (This would be called a 3×3 factorial, or 3^2 factorial.) Nine would therefore be the minimum basic number of different treatments, and we should try to arrange for an experiment with a total number of observations that was a multiple of nine. A convenient lay-out, having regard to the possibility of using distinct blocks to control variations in fertility, would be to use several blocks each with nine plots. Each block would contain plots treated by all nine possible combinations of variety and fertiliser level. This is the typical factorial arrangement, and it may easily be extended to accommodate several different factors.

There are many important advantages arising from this type of experimentation, some of which are rather surprising at first sight, as compared with the standard classical method:

(i) We obtain a broad picture of the effect of each factor in the different conditions furnished by variations in the other factors.
(ii) The use of a wide range of factor combinations provides a more reliable basis for making practical recommendations that will be valid in variable circumstances.
(iii) If all the factors happen to act independently of each other, we obtain as much information about *each* from a single experiment as we should if the whole experiment were devoted to only one factor.
(iv) If the factors are not independent of one another, we collect automatically a great deal of information about the nature of

the interaction; indeed, it is only a factorial experiment that can give a satisfactory account of this complication.

The remarkable property indicated in (iii) means that for a given number of experimental units we can often obtain a very considerable increase in efficiency merely by using a more carefully patterned design. This is still true even if the factors interact, when, as implied in (iv), a factorial arrangement is essential to making proper estimates of the type of interaction present.

The possible range of factorial designs is very large indeed, and in this chapter we shall do no more than examine the general ideas and look at one or two simple examples. It is more important that the reader should appreciate the potentialities of factorial experimentation and know how to interpret results than that he or she should personally be able to design and analyse a large number of complicated lay-outs.

However, a warning should be given against trying to include too many factors in one experiment. Although, theoretically, the inclusion of a large number leads to greater efficiency, there is often considerable difficulty in handling and interpreting very complex results.

12.2 BASIC IDEAS AND NOTATION IN THE 2^n FACTORIAL

We shall first consider designs in which there are several factors each at two levels. When there are n factors, we call this a 2^n factorial. 'Levels' may be quite literally two quantitative levels or concentrations of, say, a fertiliser, or it may mean merely two qualitative alternatives such as the different species of a plant. In some cases one level is simply the absence of the factor and the other its presence.

Suppose we indicate by the capital letters A, B, C, . . . the names of the factors involved; and by the small letters a, b, c, . . . *one* of the two levels of each of the corresponding factors. In a specific trial to test the effect of fertilisers, we might use the ordinary notation of K, N and P to indicate potash, nitrogen and phosphate; and the letters k, n and p to represent the presence of some specified concentrations. Thus the 'treatment' knp means that all types were being applied; k means potash only, the other two being absent; and the absence of all three is indicated by '1'. (Instead of 'absence', we might prefer to consider presence at a second, perhaps lower, concentration.)

With three factors A, B, and C, there are evidently eight different kinds of treatment, namely

$$\text{'1'}, a, b, ab, c, ac, bc, abc,$$

and we could construct a design using several different blocks, each of which contained exactly eight plots, one for each treatment. The yields of the plots can be subjected to an analysis of variance technique, with a final summary in an analysis of variance table and a table showing the average effects of the factors, both separately and in different combinations. We first require a more extended notation in order to present the analysis in a concise form and to explain the meaning of what is being done.

Suppose there are r blocks, or *replicates* as they are often called in this context. We write T_1, T_a, T_b, T_{ab}, etc., for the total yields of the r plots having treatments '1', a, b, ab, etc., respectively. The corresponding mean values, obtained by dividing these by r, are $\bar{1}$, $\bar{a}$, $\bar{b}$, $\overline{ab}$, etc. (Some writers use a notation in which the totals are written as [1], [a], [b], [ab], etc., and the means as '1', a, b, ab, etc., without the bars.)

We shall now define *main effects* and *interactions*. In order to do this in the simplest possible way, let us first consider an experiment with only two factors, A and B. The effect of factor A can be represented by the difference between mean yields obtained at each level. Thus the observed effect of A at the first level of B is $\bar{a} - \bar{1}$; and the observed effect of A at the second level of B is $\overline{ab} - \bar{b}$. The average observed effect of A over the two levels of B is called the *main effect*, and this is therefore defined by

$$A = \tfrac{1}{2}(\bar{a} - \bar{1} + \overline{ab} - \bar{b}),$$

where we are using the symbol A to represent the main effect of the factor A. A similar argument gives the main effect of B as

$$B = \tfrac{1}{2}(\bar{b} - \bar{1} + \overline{ab} - \bar{a}).$$

Now, if the two factors act independently of one another, we should expect the true effect of one to be the same at either level of the other. We should, for example, expect that the two observed quantities $\bar{a} - \bar{1}$ and $\overline{ab} - \bar{b}$ were really estimates of the same thing. The difference of these numbers is therefore a measure of the extent to which the factors interact, and we therefore write the *interaction* as

$$AB = \tfrac{1}{2}\{(\overline{ab} - \bar{b}) - (\bar{a} - \bar{1})\}.$$

A useful approach, as we shall see, is to assume that an interaction is non-existent unless the observed value departs significantly from zero.

These ideas are easily extended to several factors. With three, A, B and C, we have of course three main effects, which we can call A, B and C; three 'first-order' interactions, AB, AC and BC; and one 'second-order' interaction, ABC. The 'second-order' interaction is a slightly more subtle concept to grasp than the 'first-order' quantity, but it can be thought of in various ways, such as the difference in the interaction AB calculated at each of two levels of C. Similar considerations apply to higher-order interactions in experiments with larger number of factors.

12.3 THE ANALYSIS OF VARIANCE FOR A 2^n FACTORIAL

We are now in a position to discuss the analysis of variance which must contain contributions to the total sum of squares from each main effect and interaction, as well as from the blocks or replicates. When all these items are subtracted from the total sum of squares, calculated as usual, we are left with the residual sum of squares which is needed to obtain the residual variance s^2.

The actual contribution to the sum of squares for any main effect or interaction can be obtained in terms of certain complicated sums and differences of the observed yields. Fortunately, a simple computational rule enables us to avoid discussing these

Table 25. *A 2^3 factorial design, laid out in four blocks, to determine the effect of different kinds of fertiliser on potato crop yield*

Block 1

kn 284	'1' 98	*kp* 372	*k* 257
n 111	*knp* 446	*p* 302	*np* 361

Block 2

kp 385	*knp* 429	*p* 317	'1' 105
k 283	*n* 95	*np* 328	*kn* 311

p 305	*kn* 320	'1' 85	*n* 133
np 335	*k* 267	*kp* 399	*knp* 464

Block 3

np 351	*kp* 422	*k* 298	*knp* 452
n 97	'1' 126	*kn* 277	*p* 313

Block 4

explicitly, and this is illustrated in the following example. Let us consider the 2^3 design shown in Table 25. The purpose of the experiment is to determine the effect of different kinds of fertilisers, K, N and P, on potato-crop yield, and it is laid out in four replicates. Table 26 summarises the various block and treatment totals.

The first step in the analysis is to remove the block effects. To do this we simply carry out a randomised block analysis, as described in section 11.3, for the eight treatment combinations and the four blocks. The initial calculations are therefore:

Total number of observations = 32.

Grand total, $G = 9128$. Correction factor $= \frac{1}{32}G^2 = 2\,603\,762.0$.

Total sum of squares about the mean

$$= 284^2 + 98^2 + \ldots + 277^2 + 313^2 - 2\,603\,762.0$$

$$= 433\,618.0.$$

Sum of squares for blocks

$$= \tfrac{1}{8}(2231^2 + 2253^2 + 2308^2 + 2336^2) - 2\,603\,762.0$$

$$= 879.3.$$

Table 26. *Summary of block and treatment totals, and grand total*

Block total	1: 2231	Treatment total	'1': 414
	2: 2253		k: 1105
	3: 2308		n: 436
	4: 2336		kn: 1192
			p: 1237
			kp: 1578
Grand total: $G = 9128$			np: 1375
			knp: 1791

Sum of squares for treatments

$$= \tfrac{1}{4}(414^2 + 1105^2 + \ldots + 1791^2) - 2\,603\,762.0$$

$$= 426\,723.0.$$

The first part of the analysis of variance is therefore as set out in Table 27. The residual variance is $s^2 = 286.5$, based on 21 degrees of freedom. Large differences between the eight treatment combinations are clearly evident, as we should expect from inspection of the treatment totals shown in Table 26. It so happens that the variation between blocks is about equal to the residual, so that in this example there is no special contribution from fluctuations in soil fertility, and the division into blocks has not led to any gain in accuracy.

The next step is to pick out the individual items in the treatment sum of squares corresponding to the various main effects and interactions. This is where we use the special computational rule, referred to above, that enables us to avoid specific algebraic formulae for the sums of squares required. Table 28 shows the calculations we need. It is an essential part of the procedure that the treatment combinations be written down in a standard order: each factor is introduced in turn, and is then followed by all combinations of itself with the treatment combinations previously written down. Against each treatment combination we write the corresponding total yield.

We next form the entries in column 1 as follows. The first four entries are the sums of *successive pairs* of the numbers in the total

Table 27. *Analysis of variance for Table 25*

Source of variation	Sum of squares	Degrees of freedom	Mean square	Variance ratio
Blocks	879.3	3	293.1	1.02
Treatments	426 723.0	7	60 960.4	212.8
Residual	6 015.7	21	286.5	–
Total	433 618.0	31	–	–

yield column. The second four entries are the differences of these pairs, *the upper figure always being subtracted from the lower*. To obtain column 2 the whole process is repeated on column 1, and 3 is derived from 2 in a similar fashion. This 'sum-and-difference' procedure is performed as many times as there are factors, three times in fact in the present example.

The first term in column 3 is the grand total of all the observations, which should check with a straight addition. Other entries in column 3 are the totals for main effects or interactions corresponding to the treatment combinations in the initial column of the table. If we want the actual values of these factorial effects, as required by the definitions of section 11.2, then the totals must be *divided* by $r2^{n-1}$, where r is the number of blocks and n the number of factors. In the present example this divisor is $4 \times 2^2 = 16$. Thus the main effect of K, i.e. the average of all plots with k minus the average of all plots without k, is $+2204/16 = +138$. The variance of each of these factorial effects is estimated by $s^2/r2^{n-2}$. Confidence limits can then be attached in the usual way.

We can test the significance of the factorial effects directly from the totals in column 3 of Table 28. Each total has a variance of $32s^2$ and a standard error of $\sqrt{(32s^2)}$, which in this case is $\sqrt{(32 \times 286.5)} = 95.7$. In general, to obtain the standard error we merely multiply s by the square root of the total number of observations. Significance tests are now based on the t-distribution with 21 degrees of freedom. The 5 per cent points for the totals are therefore $\pm 2.080 \times 95.7 = \pm 199$, and 1 per cent points

Table 28. *Scheme of calculations for finding the treatment main effects and interactions*

Treatment combination	Total yield	(1)	(2)	(3)	Effect
'1'	414	+1519	+3147	+9128	(Grand total)
k	1105	+1628	+5981	+2204	*K*
n	436	+2815	+1447	+460	*N*
kn	1192	+3166	+757	+140	*KN*
p	1237	+691	+109	+2834	*P*
kp	1578	+756	+351	−690	*KP*
np	1375	+341	+65	+242	*NP*
knp	1791	+416	+75	+10	*KNP*

are $\pm 2.831 \times 95.7 = \pm 271$. We then see that the main effects, K, N and P, are all highly significant, being well beyond the 1 per cent point. The interaction KP is also significant at this level, while NP is between the 1 and 5 per cent points. The other two interactions are not significant.

One way of interpreting the *positive* interaction of KN is to say that potash and nitrogen do not act independently of one another; when both are present their individual effects are enhanced, although in this experiment the result is not significant. Similarly for the interaction NP. On the other hand, the KP interaction is *negative* in this experiment. So when potash and phosphate operate jointly the full benefit of each is not achieved.

A check on the treatment sum of squares in Table 27 can be made by direct calculation from Table 28. The contribution from each factorial effect is given by the square of the corresponding total divided by the number of observations. The treatment sums of squares should therefore be

$$\tfrac{1}{32}(2204^2 + 460^2 + \ldots + 242^2 + 10^2) = 426\,723,$$

checking with Table 27.

These are the basic calculations required for the analysis of simple factorial experiments of the 2^n type. A certain amount of further information can be extracted from the material by, for instance, presenting the mean yields of certain combinations of factors averaged over the rest. The actual use to which one puts the results of a factorial experiment, and the mode of interpretation adopted, depend of course on the precise context. A good deal of experience is required to grasp the full implications of a large factorial experiment, especially when some of the further complications mentioned in the next section are involved.

12.4 THE SCOPE OF MORE ADVANCED DESIGNS

Although in this chapter we have only been able to look at the general principles of factorial experimentation and the way these work out in the 2^n type of design, it is worth while giving some indication of the scope of more advanced applications. This will enable the reader to consider for herself or himself the various possiblities that might be tried in relation to any particular experimental programme. However, the actual execution of a more elaborate design will usually require some assistance from a statistical expert.

In the first place we have looked in detail at designs involving factors having only two levels, for which the analysis is relatively simple. It is quite possible to handle designs for several factors having any number of levels, and in section 11.5 we saw how the randomised block design could be used when there were just two factors, each having several levels.

The important departures from a simple 2^n factorial described in this section stem from the requirement that block sizes should be kept reasonably small if the residual variation is not to be unduly large. With a 2^3 design there are eight different treatment combinations, and so we naturally tend to use several blocks each of eight plots. Now, the fertility of such blocks is often reasonably homogeneous. But as the number of factors increases, and with it the number of plots in a block, we may be faced with serious heterogeneity. One important device for retaining a small-block size is called *confounding*. We might have a 2^5 design with 5 factors. Instead of using blocks of 32 plots, it is possible to arrange that a selection only of the full 32 treatment combinations appears in each block. We could, for example, spread the

32 treatment combinations over 4 blocks each of 8 plots. If this is done in the right way we can still obtain the information we require on main effects and first order interactions, at the cost of sacrificing information on certain higher-order interactions.

Another expedient is to use *fractional replication*. In this we use only a fraction of the full set of treatment combinations in a single block. Thus 5 drugs each at two levels could be tested in a factorial experiment on only 16 mice, using a suitably chosen $\frac{1}{2}$-replicate. Again certain information must be sacrificed, but, properly handled, such experiments can be very economical and time-saving, and are specially valuable in exploratory investigations.

Subtleties such as confounding and fractional replication, which are often combined in the same design, are most readily applied in 2^n factorials, but they can also be adopted with advantage in other designs with all factors having the same number of levels, such as 3^n, 4^n and 5^n. Difficulties are, however, liable to arise if the experiment is 'mixed' in the sense of having factors with different levels, and it is, in general, inadvisable to embark on such intricacies.

A good example of the reduction in the size of an experiment is afforded by an investigation into the consistency of certain foods. There were 7 factors involved relating to the shape and size of the cone in an Adams consistometer, the temperature, the general level of consistency in the material, etc. Assuming three basic levels for each factor, we should have a 3^7 factorial with $3^7 = 2187$ basic treatment combinations. This is an unmanageably large number, and so a $\frac{1}{9}$-replicate was devised having only 243 treatment combinations. A block in this context would be a day's work, and 243 experimental determinations was certainly too large for the size of a single block. However, it was possible to employ confounding as well, and to distribute the 243 small experiments over 9 days, with 27 determinations for each day. This reduced the scale of the initial factorial arrangement to manageable proportions, but was still sufficient to provide adequate information on main effects and first-order interactions.

These remarks must suffice to indicate the kind of advantages that can be obtained from advanced experimental designs. Their application usually requires expert statistical assistance, and if this is not available, it is best for the experimenter to use something that is simpler and more easily understood, even if it is theoretically less efficient.

At the same time it should be emphasised that the use of fully computerised methods can make an enormous difference to the handling of more advanced designs, and is also worth while in analysing the simple designs discussed above. The use of statistical software specially adapted to this type of investigation not only takes care of the management of large data sets, but also deals automatically with the tedious and lengthy computations involved.

13

Random samples and random numbers

13.1 THE NEED FOR RANDOM SELECTION

Frequently in this book we have talked about 'random' samples and 'random' selection. The word 'random' has been used in a more or less intuitive way to imply the purely haphazard and unbiased collection of measurements or allocation of treatments to the experimental material. At the same time, it has been hinted that it is often necessary to ensure that the events in question are sufficiently random by appealing to some special technique. Although one can easily get into deep philosophical waters over questions of the 'true meaning of randomness', it is unnecessary to broach such difficulties here. So far as the experimenter is concerned, he or she wants to avoid the sort of bias that will lead him or her to infer, for instance, real differences between treatments, when the observed differences are due merely to some heterogeneity in the test material. We shall therefore discuss randomness only in relation to avoiding various forms of conscious and unconscious bias. The use of tables of random numbers is the chief means of achieving this object. (It is, however, worth mentioning in passing that randomisation in properly designed experiments, such as those with randomised blocks, for example, does in fact make the usual significance tests approximately valid even when the assumption of normally distributed residual variation does not hold.)

The possibility of simple forms of bias will of course occur quite naturally to the competent experimenter. If you are testing the effect of two different types of diet on two groups of rats, then obviously the two groups of rats ought to be in some sense equivalent in all relevant features. What features are relevant can be decided only by experience and general background knowledge. It would clearly be unwise to feed animals from one inbred strain on the first diet and animals from another inbred strain on the second diet; we know that different strains may easily have different metabolic characteristics, and so an observed difference in growth-rates might be due to the effect of strain rather than diet. Similar remarks apply to other ways of classifying experimental animals, such as by age, sex, size, weight, previous diet, state of health and so on. We may be prepared to dismiss some of these factors as unimportant. There are two main ways of dealing with those that remain.

First, we can allot the whole set of rats available at random to the two treatments, without making any distinction in the factors we regard as influencing the effect of diet, and then use a t-test for comparing the mean growth-rates of the two groups. The idea is that any factor such as strain or sex will have an equal chance of falling to either treatment, and so the variation entailed by such factors is absorbed in the residual variation. This reasoning is perfectly valid. The snag is that if the animals are a rather heterogeneous collection the residual variation will be large and will therefore render the experiment insensitive to differences between the two treatments. Mention has already been made of this disadvantage in Chapter 6, where we saw how two drugs could be tested more efficiently on a set of heterogeneous subjects by applying both drugs at different times to each patient, and then examining the series of differences of pairs of corresponding responses. Similarly, in Chapter 11 it was pointed out how large natural variations could be reduced by means of a randomised block design as compared with a completely randomised arrangement.

A better plan in the dietary experiment would therefore be to divide up the rats into pairs, each of which was comparatively homogeneous. That is, the two members of each pair would be, so far as possible, similar with regard to strain, age, sex, weight, etc. We could then allot (at random to be on the safe side) the two members of each pair to the two diets, and analyse results by means of the paired-comparison test of section 6.2. With more than two diets we should require small groups each large enough to provide as many animals as there were diets to be tested. This would then lead immediately to a randomised block design, in which the diets were the treatments and the blocks were the homogeneous groups of rats. More generally, we may choose as blocks any groups of experimental units that are homogeneous with regard to characters possibly affecting the results.

A more ambitious scheme would be to attempt a factorial experiment in which age, sex, weight-class, etc., constituted the various factors. This would certainly be more informative about the general effect of diet under varying conditions, but it would also be more difficult to obtain a balanced arrangement in which all combinations of factor levels were represented.

As an alternative and with a view to keeping the design simple, we might consider using only two groups of rats, one for each of the two diets, specially chosen to match each other in the relevant factors instead of being decided at random. Thus each group would contain, so far as possible, the same proportion of animals of each strain, age, sex, weight, etc. This procedure is not to be recommended. Compared with purely random allocation the residual variation would be somewhat reduced, although, as with quota-sampling in social surveys, the proper residual variance is difficult to assess. In general, the method leading to a paired-comparison test or randomised block lay-out is much superior.

Such considerations apply equally well to any form of experimental design, whether we are allocating treatments to animals, plots of land, bacterial colonies, fruit trees, etc. It may also happen that the experimental units are not under quite such strict control. We might have to rely on making observations under specified conditions. This sometimes occurs in the study of insect or animal populations, for instance, when we want to observe behaviour against a range of meteorological or other environmental conditions. We might have a factorial 'experiment' in which the factors involved were sunshine, rain, humidity and temperature. Although we cannot go into details here, the same principles apply. We first try to imagine all reasonable sources of possible bias, and then try to choose some system of random sampling that will overcome the bias.

Suppose now that we want to take a representative sample of plants from a strongly growing field of wheat. How should this be done? By 'representative' one means, of course, a sample in which ideally all variations in quality are present in the proper proportions. In practice one must naturally be satisfied with something that is only an approximation to this. Nevertheless, one wants to avoid bias towards any particular type of plant. If, for example, we cast a hoop into the field, it would be more likely to come to rest in or near a patch of plants that were taller than average. Moreover, we should probably tend to aim somewhere near the centre of the field and away from the border where growth is likely to be poor. With practice we might hope to do better, though it has usually been found in many different spheres that even experienced observers tend to produce heavily biased samples: e.g. the plants selected are on average too tall, too short, have too many large ears, show too much or too little variation and so on. No matter how reasonable and honest one tries to be, there is always a strong likelihood of being influenced by purely subjective factors causing unconscious bias. In section 8.6 mention was made of the tendency of some laboratory technicians to obtain repeats of differential white-cell counts that were much closer to the original one than would be expected, unless special steps were taken to eliminate unconscious bias by means of a mechanical method of scoring the results.

In the case of the field of wheat we want to be able to select a number of points in it at random, perhaps collecting all plants within some specified radius of each point. With a rectangular field we shall achieve our object if we take points whose distances, measured say to the nearest metre, from two adjacent sides are selected at random.

Each individual problem may have its own special difficulties with regard to random allocation or selection, and some ingenuity may be required to choose a method that will avoid known types of bias with a reasonable economy of effort. In the next section we consider in detail the technique of using random numbers to achieve the desired result.

13.2 THE USE OF RANDOM NUMBERS

Various sources of random numbers are available, but one of the most convenient and accessible for ordinary statistical purposes is in Fisher & Yates' *Statistical Tables*. The first six pages of the table contain 15 000 numbers arranged in pairs. These numbers have been generated so as to be an effectively random or haphazard series of digits. Without embarking on any detailed logical analysis, we can loosely define what we mean by 'random' in this context as follows. Each of the digits 0, 1, 2, . . . , 8, 9 ought to appear in a long series with approximately equal frequencies; so should all possible pairs 00, 01, 02, . . . , 99, and all possible triplets, etc. In fact, any specific pattern of digits ought to appear with approximately its calculable chance of occurrence. The tables have been tested to make sure that they have the elementary properties expected of them. It is obvious, however, that if one looks carefully enough at any given series one will always discover some pattern or property which could be regarded as evidence of non-randomness. This need not worry us in ordinary biological experimentation: we only want to avoid the kinds of bias mentioned in the last section, and so long as the random numbers available will do this for us they are good enough.

The main problem is how to use an effectively random series of digits to allocate experimental units at random to a number of specific treatments. Suppose we have 45 rats on which to test three kinds of diet. If they are all fairly similar in age, sex, weight, etc., we may decide that the best thing is simply to divide them at random into three groups. We could do this by earmarking the first 15 that came to hand (either physically as we moved from cage to cage, or theoretically by going through a litter-book) for treatment by the first diet, the second 15 for the second diet, and the rest for the third diet. Such a procedure is always liable to be somewhat suspect even if we cannot see any possibility of bias. Some unconscious or environmental bias might creep in with regard to the cages we approached first, or the order in the litter-book might be related to some unsuspected genetical trend. Moreover, hostile critics could always assert the possibility of bias, and might even discover it after the experiment was over.

The best plan is to label the animals with the number 1 to 45; then use the table to arrange these numbers in random order. We can then assign the first 15 rats to the first treatment, the second 15 to the second treatment and the rest to the third treatment. In this way we shall avoid the risk of subsequent allegations of hidden bias.

To carry out the randomisation we first read across the lines of random numbers writing down those that are 45 or less and rejecting 00 and any repetitions. Starting at the beginning of Fisher & Yates' table, we obtain 03, 43, 36, 33, 26, 16, . . . , having rejected numbers like 47, 73, 86, . . . because they are greater than 45, and also ignoring 36 when repeated after 47. The required random order therefore starts off with 3, 43, 36, 33, 26, 16, . . . , and the rats with the first 15 identification numbers to appear are all allotted to the first treatment. Rats with the next 15 numbers are assigned to the second treatment and the remainder to the third treatment.

The above process entails rejecting, on average, 55 numbers in every 100 selected from the table, not counting repetitions, namely 46–99 and 00. Again, if we were arranging 30 numbers in random order we should be rejecting 70 in every 100. This is an unnecessary waste of time, as most of the rejected numbers can be brought into service as follows. Take the case of the 45 rats. Suppose we accept every number thrown up by the table at face value if it is 45 or less, and subtract 50 from it if it is greater than 50. The numbers 51–95 would then yield 1–45, and the only numbers in the table which we should have to reject would be 00, 46–50 and 96–99, i.e. only 10 in every 100.

If we had, say, 35 rats to be put in random order, we could subtract 40 from numbers over 40. This would still mean rejecting 30 in every 100. It might be thought that we could use numbers over 80 by subtracting 80 ($= 2 \times 40$) from them. This would not, however, be correct, because 81–99 would yield only 1–19, and the remaining numbers 20–35 would not be supplied from this source. Thus, all numbers between 1 and 35 would not have an equal chance of being chosen to fill a given place in the final series; we should be more likely to find numbers between 1 and 19 towards the beginning.

There are many modifications of the above ideas that can be used to suit particular circumstances, but it is better to use a simple method known to be correct rather than a more elaborate procedure which, though quicker, may introduce some unsuspected bias.

In order to aid the random arrangement of small sets of numbers, Fisher & Yates also give random permutations of the 10 numbers 0–9, and random permutations of the 20 numbers 0–19. These tables avoid some of the difficulties mentioned above, such as large random numbers having to be modified by subtraction of a suitable constant or the rejection of repetitions.

In a simple factorial lay-out of type 2^3 there are eight basic treatment combinations per block, i.e. '1', *a*, *b*, *ab*, *c*, *ac*, *bc* and *abc*. If we are applying this to an agricultural experiment, each area of land making up a block has to be divided into eight plots. To avoid bias it is advisable to allot the eight treatments at random to the eight plots of each block, using different randomisations for the different blocks.

Suppose we label the eight treatments, as written in the order shown above, with the numbers 1 to 8. We then take a random permutation from the tables, dropping out 0 and 9. The first one given in the tables thus yields 3, 5, 2, 8, 4, 7, 6, 1. Taking the plots of the first block in some systematic order, we therefore assign to them the treatments *b*, *c*, *a*, *abc*, *ab*, *bc*, *ac* and '1'. For the next block we take a new random permutation and so on.

If we are carrying out only an occasional experiment requiring some deliberate randomisation, we can start reading the numbers from any haphazardly chosen point or can take a permutation from anywhere on a page. But when a series of experiments is involved we should be more careful, and should work through a section of the table systematically so as to avoid any sequence being used twice (except in so far as it may reappear by chance in the random numbers themselves).

Finally, it should be mentioned that random numbers can also be generated *ad hoc* by some pocket calculators and also by some computer software. But this can introduce additional complications and is generally not to be preferred to the use of tables as described above unless very large amounts of such numbers are required and need to be manipulated automatically (in addition to being calculated).

13.3 SURVEYS AND CENSUSES

The subject of surveys is one in which problems of random sampling loom very large. This topic is a rather specialised one, and it would take us too far afield to discuss it in any detail. Nevertheless, it is worth while drawing attention to some of the main uses of surveys in biology and medicine, and pointing out the importance of obtaining adequately representative samples of data by means of an appropriate method of randomisation.

Some surveys are intended more for the preliminary investigation of a problem rather than for the precise testing of specific hypotheses. Suppose we want to make an ecological study of a particular geographical region and plan some preliminary inquiries to find out the main features of the area. A fairly cursory inspection may show that there is considerable variety in the landscape: moorland, bog, grassland, rivers, and so on. Each of these items may appear in several distinct types. Initially, at any rate, all these various aspects of the terrain may appear to have roughly equal claims to one's attention. In this case we should consider a small pilot survey in which approximately equal amounts of information were collected from each main aspect. Later we may decide that some aspects are unimportant and will therefore concentrate more attention on the remainder.

Having reduced the main features of the region to a number of more or less homogeneous items, we are faced with the problem of surveying each of these in more detail. An area of woodland or grassland can perhaps be divided up on a map into a number of small squares. We then select squares at random for individual investigation. Using the methods of the last section, we simply number the squares in some standard order and then refer to a table of random numbers. If there were 15 squares in a particular area and we thought that we had time to examine only 3, then the first 3 pairs of random numbers (lying between 1 and 15) obtained from the table would tell us which squares to look at in detail.

A similar technique can easily be applied in other contexts. If we want to make a broad clinical survey of a particular disease, we may prefer to start by selecting, say, 10 patients at random from each sex and age-group, the latter being in 5- or 10-year intervals. We should do this even if the majority of cases occurred in men between 45 and 50. The reason is that for scientific purposes we want to obtain data over a good range of possibly relevant conditions. In the study of a disease we should probably want to extend the basic classification to include the Registrar-General's main social classes, the distinction between town and country, etc.

Obviously there is a limit to the extent to which we can introduce additional classifications and still obtain sufficient data in each sub-class. This is largely a question of judgment. The main point being made here is the importance of giving adequate weight to relatively uncommon combinations of factors, and not allowing these to be swamped by excessive quantities of information about the commoner situations.

If, on the other hand, we were primarily concerned with a census rather than a survey, then a different attitude would be adopted. In a census we want to obtain an inventory of the material in question, say the timber in a forest. Since we are unlikely to be able to catalogue each tree separately, we must adopt a suitable sampling scheme. But now we want the sample to reflect the character of the forest in a representative way. The main subdivisions of the forest should appear in the sample in the same proportions. A small group of larch trees in an oak forest may be of no economic consequence, and would hardly be represented at all in, say, a 1 per cent sample. To the ecologist, however, the scientific importance might be considerable, and his survey might well pay special attention to the larch trees.

These brief remarks must suffice to indicate the kind of sampling problems involved in surveys and censuses. Rather difficult statistical problems are frequently encountered in the design, analysis and interpretation of such inquiries, and it is nearly always advisable to seek expert advice before starting on such investigations.

14

Partial correlation and multiple regression

14.1 INTRODUCTION

In Chapters 9 and 10 we saw how the techniques of correlation and regression could be used to study the association between two quantitative variables. Which method was most appropriate depended to a great extent on the sort of mathematical picture that best described the data. As we saw, correlation techniques were suitable when we had two measurements that approximately followed a bivariate normal distribution. If, on the other hand, one measurement, x, was arbitrarily or irregularly distributed, and we were chiefly interested in the behaviour of a second measurement, y, in relation to the first, then regression might be more appropriate. The basic assumptions made in that case were that the true means μ_x, when plotted against the corresponding values of x, lay on a straight line, and that variation about this line followed a normal distribution whose variance was independent of x.

These methods are extremely powerful when we have only a single pair of variables to investigate. Quite often, however, measurements of different kinds will occur in batches. Although we could use the simple correlation or regression analyses, taking the variables two at a time, this would not only be unwieldy but would fail to reveal more complicated patterns of relationship that might exist between several variables simultaneously. We therefore require extensions of the previous methods specially adapted to handling groups of variables, and the appropriate techniques are those of *partial correlation* and *multiple regression*. With suitable assumptions these enable us to disentangle the mutual dependence of a set of variables and to show how any pair are related when the effect of the others has been eliminated. We first consider the problem of partial correlation.

14.2 PARTIAL CORRELATION

It was pointed out at the end of section 9.4 that the existence of a correlation between two factors, A and B, does not in itself necessarily imply a direct causal link. Factor A might of course cause B, or B might cause A. Again, some third factor, C, which influences both A and B, might be responsible for their mutual correlation. Under certain conditions the effect of diet on both height and weight illustrates this sort of thing. As a rule, the correlation points to some kind of link, though it does not automatically specify the nature of the relationship. (On occasion entirely spurious correlations can arise, but this is usually due to not observing the conditions required for the analysis to be valid.) It follows, therefore, that when several variables are involved the study of suitably calculated correlation coefficients may provide important clues to the underlying causal mechanism.

Let us suppose that we wish to investigate the three factors A, B and C, which give rise to measurements x_1, x_2 and x_3 respectively. Using the methods of Chapter 9, we can calculate an estimate r_{12} of the true correlation coefficient ρ_{12} between A and B. Similarly, the estimated and true coefficients are r_{13} and ρ_{13} for A and C; and r_{23} and ρ_{23} for B and C. Since correlation is a symmetrical relationship, it clearly does not matter in what order we write the numerical suffixes, e.g. $\rho_{12} = \rho_{21}$, etc. All these coefficients may be designated 'total' in order to distinguish them from the 'partial' coefficients defined below.

If we were primarily interested in the correlation between A

and B, irrespective of any influence C might have, we should concentrate attention on the estimator r_{12}. But we might, on the other hand, wish to know how A was related to B when the effect of C had been specifically excluded, and, conversely, how A was related to C when B was excluded. For this purpose we require *partial correlation coefficients*. The partial correlation coefficient of A and B, excluding C, is represented by $\rho_{12.3}$. The general rule in this notation is that the pair of suffix numbers before the point relates to the variables under comparison, while numbers after the point correspond to variables specifically excluded from the comparison; there may, in general, be still other numbers not mentioned in the suffix which represent variables about which no particular decision has been taken.

Now any partial correlation coefficient can always be expressed in terms of three other coefficients having one fewer numbers in the suffix after the point. The basic formula required exhibits the estimate $r_{12.3}$ in terms of r_{12}, r_{13} and r_{23}, and is

$$r_{12.3} = \frac{r_{12} - r_{13}r_{23}}{\sqrt{[(1 - r_{13}^2)(1 - r_{23}^2)]}}. \tag{67}$$

Similar formulae for $r_{13.2}$ and $r_{23.1}$ are easily obtained by interchanging the numbers. We also have analogous expressions for the true partial correlation coefficients $\rho_{12.3}$, $\rho_{13.2}$ and $\rho_{23.1}$, in which every r in formula (67) is replaced by a ρ.

If there were four variables A, B, C and D, represented by 1, 2, 3 and 4 respectively, we could carry the process a stage further and define, for instance, $r_{12.34}$. This is an estimate of the correlation between A and B when C and D have been eliminated. One convenient formula is

$$r_{12.34} = \frac{r_{12.4} - r_{13.4}r_{23.4}}{\sqrt{[(1 - r^2_{13.4})(1 - r^2_{23.4})]}}, \tag{68}$$

which is simply formula (67) with a '4' inserted in all the suffixes as an additional eliminated variable. Since $r_{12.34} = r_{12.43}$, we could also express $r_{12.34}$ in terms of $r_{12.3}$, $r_{14.3}$ and $r_{24.3}$, by using formula (68) with '3' and '4' interchanged.

Formula (68) has three partial correlation coefficients on its right-hand side, each of which can be put in terms of three of the total coefficients r_{12}, r_{13}, r_{14}, r_{23}, r_{24} and r_{34} by means of formulae like formula (67). It should now be clear how we can calculate successively any partial coefficients that may be required, starting with the total coefficients whose computation was described in Chapter 9.

Partial correlation coefficients can be tested for significance in much the same way as total correlation coefficients, and we simply use the earlier formulae (40), (41) and (42) as though they referred to partial instead of total coefficients. However, an important adjustment is required to allow for the number of variables eliminated. Formula (40) is unaffected since it applies only to very large samples. But in formula (41), and in the variance $1/(n-3)$ associated with formula (42), we replace n by n minus as many variables as have been eliminated from the comparison in question.

Also worth mentioning is the validity of the usual significance tests in connection with partial correlation coefficients. With only two variables, we assume the basic joint distribution to be bivariate normal. There is a natural extension to the case of larger numbers of variables which leads to a *multivariate normal distribution*. The chance of this distribution being approximately followed in practice is much smaller than the chance of a bivariate normal distribution being adequate for a pair of variables. However, the procedures suggested above for testing partial correlation coefficients do not depend on the distribution of the eliminated variables. The latter may therefore have any distribution or indeed be quite arbitrary, and the more restrictive assumptions are required only for the set of variables retained for special consideration.

In warning against possible misinterpretations of the correlation coefficient at the end of section 9.4, it was pointed out that height and weight, for example, might be positively associated because both depended on the general process of growth. If we wanted to know whether height and weight were still positively correlated when the effect of growth was allowed for, we might try the methods of partial correlation. Let us consider data for a group of schoolboys. Growth, as such, is not easy to measure directly, and in the first instance we could try using age as a rough indication. In principle we could examine the correlation of height and weight in each separate age-group, but if the numbers available were small we should be obliged to treat the sample as a whole.

Suppose we have n (= 80) boys and record for each his height, weight and age, representing these variables by the numbers '1', '2' and '3'. Using the methods of Chapter 9, we first find estimates of the total correlation coefficients. Let these be

$$r_{12} = +0.82,\ r_{13} = +0.75,\ r_{23} = +0.86.$$

All these values are fairly high as they stand, but we want to know what happens when variable 3 is eliminated from the reckoning. To do this we use formula (67) to obtain $r_{12.3}$, i.e.

$$r_{12.3} = \frac{+0.8200 - 0.6450}{\sqrt{(0.4375 \times 0.2604)}} = +0.5185.$$

The partial correlation of height and weight is thus estimated to be about +0.52. This is still quite appreciable though considerably less than the total value of +0.82, which we had before eliminating the effect of age. With $n = 80$ we should test the significance of an ordinary correlation coefficient using the t in formula (41) with 78 degrees of freedom. When one variable is eliminated we should put $n = 79$, and the degrees of freedom are reduced to 77, which is quite large enough for t to be nearly normally distributed. Writing $r = 0.52$ and $n = 79$ in formula (41) gives $t = 5.34$, which is highly significant. Direct reference to Appendix 4 shows that coefficients numerically greater than about 0.22 are significant at the 5 per cent level.

Some thought should be given to the validity of the above test. It is quite reasonable to assume something like a bivariate normal distribution of height and weight. But the ages of a group of schoolboys will, in general, have nothing remotely resembling a normal distribution; in all probability there will be approximately equal numbers in each age-group. However, it is precisely this variable which has been eliminated in calculating the partial correlation coefficient $r_{12.3}$. We may, therefore, accept the result of the significance test with some confidence.

14.3 MULTIPLE REGRESSION WITH TWO INDEPENDENT VARIABLES

As previously explained, regression techniques are frequently more appropriate than correlation methods. They involve fewer basic assumptions, and are generally more widely applicable. We have seen in the previous section how the correlation analysis of Chapter 9 for only two variables can be extended to cover several variables. The next task is, therefore, to consider the analogous extension of the regression methods described in Chapter 10.

The basic ideas of regression for two variables entailed taking one of them, x, as being 'independent' and having some possibly irregular or arbitrary distribution, while for each value of x the other 'dependent' variable y had a normal distribution with mean μ_x and variance σ^2 independent of x. If the true means μ_x lay on a straight line when plotted against x, the regression was called 'linear'. This is the simplest and most commonly used mathematical picture.

In the illustration used in Chapter 10 it was natural to take the average tail length of the various *Zosterops* species as the y-variable and altitude as the x-variable. This is because we wished to talk about the probable effect on average tail length of specified changes in altitude (but not vice versa). The original investigation took account of two further meteorological indices: the average maximum temperature of the three hottest months and the average minimum temperature of the three coldest months. We shall call these simply 'maximum temperature' and 'minimum temperature', represented by x_2 and x_3 respectively, measured in degrees Celsius. Altitude will be indicated by x_1, measured as before in units of 1000 m.

For the moment we shall confine attention to the pair of independent variables x_1 and x_2. The extended concept of regression that we are now going to use assumes that for any given *pair* of values, x_1 and x_2, the y measurements have a normal distribution about some true mean (say μ_{x1}, μ_{x2}) with variance σ^2 independent of both x_1 and x_2; and that the true y-means can be written in terms of x_1 and x_2 by the equation

$$y = \alpha + \beta_1 x_1 + \beta_2 x_2. \qquad (69)$$

This is the extended form of formula (43). Whereas we previously had a regression 'line', we now have a regression 'plane' if we wish to think in geometrical terms.

The quantities β_1 and β_2 in formula (69) are the *partial regression coefficients* of y on x_1 and y on x_2 respectively. They are sometimes written $\beta_{01.2}$ and $\beta_{02.1}$, in conformity with the notation already used for partial correlation, using 0 to represent y. The figure (or figures if there are more than two independent variables) after the point in the suffix indicates the variable eliminated from the effect in question. Thus, β_1, or $\beta_{01.2}$, represents the rate of increase of y relative to x_1 while x_2 is held constant.

We are now faced with the problem of estimating the constants α, β_1 and β_2 on the right-hand side of formula (69) by means of quantities a, b_1 and b_2, obtained from the data available to us. Corresponding to formula (46) we have the fitted regression given by

$$y = a + b_1x_1 + b_2x_2. \tag{70}$$

The main labour is in calculating b_1 and b_2; when this has been done we easily find a from

$$a = \bar{y} - b_1\bar{x}_1 - b_2\bar{x}_2. \tag{71}$$

As with simple regression for two variables, the chief calculations here are based on sums, and sums of squares and products. In a sample of n observations, each observation now really consists of a set of three associated numbers y, x_1 and x_2. We first calculate from the data the basic quantities

$$\left.\begin{array}{llll} n, & \sum x_1, & \sum x_2, & \sum y, \\ & \sum x_1^2, & \sum x_1x_2, & \sum x_1y, \\ & & \sum x_2^2, & \sum x_2y, \\ & & & \sum y^2. \end{array}\right\} \tag{72}$$

The array in formula (72) is the extended form of formula (35). To obtain sums of squares and products about the means we use formulae similar to formula (36), namely

$$\left.\begin{aligned}\sum(x_1-\bar{x}_1)^2&=\sum x_1{}^2-\frac{1}{n}\left(\sum x_1\right)^2=A,\\ \sum(x_1-\bar{x}_1)(x_2-\bar{x}_2)&=\sum x_1x_2-\frac{1}{n}\left(\sum x_1\right)\left(\sum x_2\right)=B,\\ \sum(x_2-\bar{x}_2)^2&=\sum x_2{}^2-\frac{1}{n}\left(\sum x_2\right)^2=C,\\ \sum(x_1-\bar{x}_1)(y-\bar{y})&=\sum x_1y-\frac{1}{n}\left(\sum x_1\right)\left(\sum y\right)=D,\\ \sum(x_2-\bar{x}_2)(y-\bar{y})&=\sum x_2y-\frac{1}{n}\left(\sum x_2\right)\left(\sum y\right)=E,\\ \sum(y_1-\bar{y})^2&=\sum y^2-\frac{1}{n}\left(\sum y\right)^2=S.\end{aligned}\right\}\tag{73}$$

The sums of squares and products about means values in formula (73) have been labelled, A, B, . . . , etc., for convenience. For purposes of computation it is advantageous to write the correction factors in the form of an array, i.e.

$$\left.\begin{matrix}\frac{1}{n}\left(\sum x_1\right)^2, & \frac{1}{n}\left(\sum x_1\right)\left(\sum x_2\right), & \frac{1}{n}\left(\sum x_1\right)\left(\sum y\right),\\ & \frac{1}{n}\left(\sum x_2\right)^2, & \frac{1}{n}\left(\sum x_2\right)\left(\sum y\right),\\ & & \frac{1}{n}\left(\sum y\right)^2.\end{matrix}\right\}\tag{74}$$

We then subtract each of these quantities from the corresponding ones in the last three lines of formula (72) to give

$$\left.\begin{matrix}A, & B, & D,\\ & C, & E,\\ & & S.\end{matrix}\right\}\tag{75}$$

The required estimates of the partial regression coefficients are now

$$\left.\begin{aligned} b_1 &= \frac{CD - BE}{AC - B^2}, \\ \text{and} \qquad b_2 &= \frac{AE - BD}{AC - B^2}. \end{aligned}\right\} \tag{76}$$

The quantity σ^2, which is the variance of y for any given fixed values of x_1 and x_2, is estimated by

$$s^2 = \frac{S - b_1 D - b_2 E}{n - 3}. \tag{77}$$

Finally, we want to know the standard errors of the estimated regression coefficients. These are

$$\left.\begin{aligned} s_{b_1} &= \frac{s}{\sqrt{\left(\dfrac{AC - B^2}{C}\right)}}, \\ \text{and} \qquad s_{b_2} &= \frac{s}{\sqrt{\left(\dfrac{AC - B^2}{A}\right)}}. \end{aligned}\right\} \tag{78}$$

Significance tests for b_1 and b_2 may be performed as before according to the t-test indicated by formula (49), the number of degrees of freedom now being $n - 3$. If neither coefficient is significantly different from zero, it may usually be assumed that the data provide little evidence in favour of an association between y and either of the other two variables, x_1 and x_2 (though exceptions are possible). If, however, one coefficient, say x_1, is significant and the other, say x_2, is not, the best plan is to drop x_2 and re-calculate the regression of y on x_1 as in Chapter 10. There is little difficulty in doing this, since all the basic sums, and sums of squares and products, are already available. No fresh computations with the original data are therefore required.

Let us now return to the problem of the association of tail length in *Zosterops* with altitude and maximum temperature. Only the chief features of the calculations need be illustrated. The total number of observations is $n = 72$. In section 10.2 we have already given the mean values $\bar{y} = 41.69$ and $\bar{x}_1 = 1.338$, together with certain sums of squares and products about means, namely

$$S = 1526.30, \quad A = 35.136, \quad D = 133.80.$$

The additional quantities $\bar{x}_2$, B, C and E were found from the data to be

$$\bar{x}_2 = 28.39, \quad B = -153.35, \quad C = 1100.59, \quad E = -802.00.$$

Using formula (76) we obtain the estimates $b_1 = +1.6017$ and $b_2 = -0.5055$. Substitution in formula (71) gives

$$a = 41.69 - (1.6017)(1.338) - (-0.5055)(28.39) = 53.90.$$

The fitted regression equation is thus

$$y = 53.90 + 1.6017x_1 - 0.5055x_2.$$

We next calculate the residual variance from formula (77). This is $s^2 = 13.14$, giving the standard deviation $s = 3.625$. The standard errors of the two partial regression coefficients are now derived by substitution in formula (78), so that we can then write the two coefficients with standard errors attached as

$$\left.\begin{aligned} b_1 &= +1.602 \pm 0.977, \\ b_2 &= -0.506 \pm 0.175. \end{aligned}\right\}$$

Comparison of the two regression coefficients with their standard errors shows that b_2 (maximum temperature) is strongly significant, but b_1 (altitude) is not significant at all. This result is rather curious at first sight when contrasted with the analysis of Chapter 10, where altitude taken by itself was found to be highly significant, having a regression coefficient of 3.808 ± 0.643. Actually, altitude and maximum temperature have a strong negative correlation ($r_{12} = -0.78$), and we are in a way merely replacing one variable by another closely associated with it. Thus, if we concentrate attention on populations with a given maximum temperature, the additional effect of altitude is not significant.

It is also interesting to examine the residual variances in the two cases. In Chapter 10, where we considered only altitude, the residual variance was 14.5, while the variance of mean tail length taken by itself without allowing for altitude was 21.5. A fair reduction in variation resulted from using the regression on altitude. In the present case, taking cognizance of both altitude and maximum temperature, we find a residual variance of 13.1. This is only slightly less than 14.5, and corresponds to the fact that the additional information gained is small. Another point worth noting is that the partial regression coefficient for altitude has a fairly large standard error: if we could reduce the residual variation, a significant altitude effect might emerge. In the next section we shall consider the case of regression with more than two independent variables, and shall extend the *Zosterops* illustration by introducing minimum temperature as an extra variable.

14.4 MULTIPLE REGRESSION WITH MORE THAN TWO INDEPENDENT VARIABLES

In many problems there may be several independent variables that might be important. The *Zosterops* analysis of the last section took account of only two, namely altitude and maximum temperature, although minimum temperature records had also been collected because there were a priori reasons for expecting this factor to have some influence. Basic formulae such as formulae (76), (77) and (78) are comparatively simple when there are only two x-variables. But as the number increases some more general procedure must be adopted.

Suppose, for the moment, we have three independent variables x_1, x_2 and x_3. This will make description easier, but will not prevent anyone who wishes extending the principles to four or more variables. There are n 'observations', each of which is a set of numbers (y, x_1, x_2, x_3). The basic calculations with these figures involve the derivation of *all* sums and sums of squares and products.

The true regression equation is

$$y = \alpha + \beta_1 x_1 + \beta_2 x_2 + \beta_3 x_3, \tag{79}$$

and the fitted equation, for which we are looking, is

$$y = a + b_1 x_1 + b_2 x_2 + b_3 x_3, \tag{80}$$

where

$$a = \bar{y} - b_1\bar{x}_1 - b_2\bar{x}_2 - b_3\bar{x}_3. \tag{81}$$

As before, the main problem is to find the partial regression coefficients b_1, b_2 and b_3.

Suppose we write, in an obvious and easily extended notation,

$$\left.\begin{aligned} \sum(y-\bar{y})^2 &= c_{00}, \\ \sum(x_i-\bar{x}_i)(y-\bar{y}) &= c_{0i};\ i = 1, 2, 3, \\ \sum(x_i-\bar{x}_i)(x_j-\bar{x}_j) &= c_{ij};\ i, j = 1, 2, 3, \end{aligned}\right\} \tag{82}$$

calculating these expressions as for the case of only two independent variables. We then consider the *three sets* of equations in X_1, X_2 and X_3, given by associating the left-hand side of formula (83) below with each of the alternative right-hand sides in turn in the order shown:

$$\left.\begin{aligned} c_{11}X_1 + c_{12}X_2 + c_{13}X_3 &= 1 \\ c_{12}X_1 + c_{22}X_2 + c_{23}X_3 &= 0 \\ c_{13}X_1 + c_{23}X_2 + c_{33}X_3 &= 0 \end{aligned}\right\}, \quad \left.\begin{matrix} 0 \\ 1 \\ 0 \end{matrix}\right\}, \quad \left.\begin{matrix} 0 \\ 0 \\ 1 \end{matrix}\right\}. \tag{83}$$

Let the three sets of solutions be

$$\left.\begin{aligned} X_1 &= a_{11} \\ X_2 &= a_{12} \\ X_3 &= a_{13} \end{aligned}\right\}, \quad \left.\begin{matrix} a_{12} \\ a_{22} \\ a_{23} \end{matrix}\right\}, \quad \left.\begin{matrix} a_{13} \\ a_{23} \\ a_{33} \end{matrix}\right\}. \tag{84}$$

The required partial regression coefficients are now found from

$$\left.\begin{aligned} b_1 &= a_{11}c_{01} + a_{12}c_{02} + a_{13}c_{03}, \\ b_2 &= a_{12}c_{01} + a_{22}c_{02} + a_{23}c_{03}, \\ b_3 &= a_{13}c_{01} + a_{23}c_{02} + a_{33}c_{03}, \end{aligned}\right\} \tag{85}$$

and the residual variance is estimated by

$$s^2 = \frac{c_{00} - b_1 c_{01} - b_2 c_{02} - b_3 c_{03}}{n - 4}. \tag{86}$$

This time we can write the standard errors of the regression coefficients as

$$s_{b_1} = s\sqrt{a_{11}},\ s_{b_2} = s\sqrt{a_{22}},\ s_{b_3} = s\sqrt{a_{33}}. \tag{87}$$

In applications to larger numbers of independent variables, say k in all, there will be further sums of squares and products in formula (82); formula (83) will have k sets of k equations in k unknowns; formula (84) will be extended to k rows and k columns, and formula (85) to k rows; while formula (86) will be modified by having k negative terms in the numerator and a denominator of $n - k - 1$, the latter being the appropriate number of degrees of freedom for t-tests.

As an example, we can consider the addition of 'minimum temperature' to the previous illustration in which we were relating the tail length of *Zosterops* to climatic factors. This additional measurement x_3 has mean $\bar{x}_3 = 11.87$. We also want several sums of squares and products not already given in the previous section. These are

$$c_{03} = -1316.28$$

$$c_{13} = -138.32,\ c_{23} = +927.69,\ c_{33} = +1658.83.$$

Formula (83) is thus

$$\left.\begin{aligned} +35.136X_1 - 153.35X_2 - 138.32X_3 &= 1 \\ -153.35X_1 + 1100.59X_2 + 927.69X_3 &= 0 \\ -138.32X_1 + 927.69X_2 + 1658.83X_3 &= 0 \end{aligned}\right\},\quad \left.\begin{aligned} 0 \\ 1 \\ 0 \end{aligned}\right\},\quad \left.\begin{aligned} 0 \\ 0 \\ 1 \end{aligned}\right\},$$

with solutions to seven decimal places

$$\left.\begin{aligned} X_1 &= +0.0731237 \\ X_2 &= +0.0095518 \\ X_3 &= +0.0007556 \end{aligned}\right\}, \quad \left.\begin{aligned} &+0.0095518 \\ &+0.0029665 \\ &-0.0008626 \end{aligned}\right\}, \quad \left.\begin{aligned} &+0.0007556 \\ &-0.0008626 \\ &+0.0011482 \end{aligned}\right\}.$$

Using formula (85) then gives the three partial regression coefficients

$$\left.\begin{aligned} b_1 &= +1.1288 \\ b_2 &= +0.0343 \\ b_3 &= -0.7184 \end{aligned}\right\}.$$

Next, going back to formula (81), we have

$$\begin{aligned} a &= 41.69 - (1.1288)(1.338) - (0.0343)(28.39) \\ &\quad - (-0.7184)(11.87) \\ &= 47.73, \end{aligned}$$

and so the full regression equation is

$$y = 47.73 + 1.1288x_1 + 0.0343x_2 - 0.7184x_3.$$

We now find the residual variance as

$$\begin{aligned} s^2 &= \frac{1}{68}\{1526.30 - (1.1288)(133.80) - (0.0343)(-802.00) \\ &\quad - (-0.7184)(-1316.28)\} \\ &= 6.723. \end{aligned}$$

Finally, using formula (87), we can write the regression coefficients with their standard errors, to three places of decimals, as

$$\left.\begin{aligned} b_1 &= +1.129 \pm 0.701, \\ b_2 &= +0.034 \pm 0.141, \\ b_3 &= -0.718 \pm 0.088. \end{aligned}\right\}$$

The hardest part of these computations is of course solving the three sets of simultaneous equations in formula (83). This is a little tedious with only three independent variables and increasingly awkward with four or more.

For those who are familiar with some elementary matrix algebra the situation is easily manageable. Suppose that the matrix of coefficients on the left of formula (83), as defined by formula (82), is represented by $\{c_{ij}\} = \mathbf{C}$; and the matrix on the right of formula (84) is given by $\{a_{ij}\} = \mathbf{A}$. Then $\mathbf{A} = \mathbf{C}^{-1}$. Many good pocket calculators have a built-in facility for the automatic solving of sets of linear equations, as well as providing elementary matrix manipulation, including inversion. So the calculation of $\mathbf{C}^{-1}$ becomes a simple matter: but make sure that the machine can do it for a reasonable range of values from two independent variables – say, up to at least six!

The column vector $\mathbf{b} = \{b_i\}$ on the left of formula (85) is then found from $\mathbf{b} = \mathbf{Ay}$, where $\mathbf{y}$ is the column vector $\{c_{0i}\}$ defined in formula (82). The residual variance about the regression line is

$$s^2 = \frac{c_{00} - \mathbf{b}'\mathbf{y}}{n - k - 1},$$

where $\mathbf{b}'$ is the horizontal transpose of the column vector $\mathbf{b}$. Standard errors are indicated as before in formula (87).

The above analysis illustrates the general calculations required for this kind of regression analysis. However, if we look at the actual values of the partial regression coefficients, we see that both b_1 (altitude) and b_2 (maximum temperature) are non-significant, while b_3 (minimum temperature) has a strongly significant influence. There is considerable reduction in residual variance, which is now only 6.7, compared with 13.1 at the end of the last section.

We can, therefore, now interpret the data most economically in terms of minimum temperature. When this is taken into account neither altitude nor maximum temperature has anything further to add by way of explaining the observed variation. The proper course is now to recalculate the partial regression equation in terms of x_3 alone, using the method of section 10.2. This may be left to the reader as an exercise.

To conclude, it must be emphasised that multiple regression with several independent variables is another area that can be very difficult to handle properly, especially with regard to the interpretation of results. The use of a computerised approach is strongly recommended, but some care is needed in choosing the right software. In any case it is advisable to seek expert assistance from a professional biostatistician.

15

Non-parametric and distribution-free tests

15.1 INTRODUCTION

In most of the preceding discussions, which involved the performance of significance tests or the estimation of *parameters* such as means, standard deviations or correlation coefficients, an essential part of the undertaking was to start by making assumptions about the statistical distributions underlying the observations. Thus, the distribution of stature in Table 1 could be assumed to be, at least approximately, Gaussian or 'normal'. Albinotic children in marriages between heterozygous partners would be expected to show a binomial distribution, as in Table 2. In Table 3 the Poisson distribution is used to describe the variable presentation of yeast cells in a haemocytometer. Later on, in discussing the correlation exhibited by pairs of measurements, as in Table 14, the more sophisticated notion of a *bivariate* normal distribution was introduced.

Of course, we do not expect all these theoretical distributions to be exactly reproduced in nature. We simply hope that in any specific situation we can choose a suitable and convenient distribution that is sufficiently accurate for practical purposes. The more sophisticated the theoretical concepts, the more trouble we shall have in justifying their practical application. Thus, in section 9.4, special consideration was given to the question whether a bivariate normal distribution for paired measurements could really be acceptable as a basis for correlational studies. The problem became even more acute when there were three or more factors, with partial correlation coefficients being introduced and the necessity of appealing to the notion of a *multivariate* normal distribution.

It is perfectly reasonable, and a part of the scientific method, to make such assumptions in order to achieve progress. But any results obtained are never absolute; they are always provisional and subject to revision. We should therefore constantly review the possibility of the assumptions being insufficiently accurate or even downright invalid. In some cases, e.g. as in sections 10.1 and 11.4, some suitable transformation of the scale of measurement may improve matters. Or again, the whole basic model may need to be recast, as in the use of a regression approach (see section 10.1) instead of a correlation method to avoid the often quite untenable assumption of bivariate normality.

However, we often feel that the underlying assumptions, e.g. that the observations are normally distributed, are sufficiently accurate for practical purposes. This means we believe that any results obtained are unlikely to be seriously affected by mild departures from normality. Another way of putting this is to say that the methods used are *robust*, i.e. in this example are not very sensitive to departures from normality. Naturally this raises many thorny questions about what departures from normality are likely to arise in practice, how these are related to changes in results obtained, and what effect this would have on overall conclusions. Without going into technical details here, it should suffice to say that, in general, when a particular approach has been shown on both theoretical and practical grounds to have a fair degree of robustness, it can be used in practice with a greater degree of confidence than an approach which is known to be highly sensitive to the initial assumptions.

A relatively extreme situation occurs when it is possible to use methods of testing significance which do not depend *at all* on the distribution of the observations. Such tests are called *distribution-*

free tests, in contradistinction to most of the tests introduced so far in this book, where the statistical properties of the test-statistics do depend in detail on the basic distribution of the observations. These latter, more familiar, types of tests are often referred to as *parametric tests*.

A this point there is, however, some confusion of terminology. Thus, certain authorities use the term *non-parametric tests* to apply where the test-statistics are independent of the parameters in the underlying distributions of the observations, but do depend on the general form, i.e. they are only parameter-free.

As an example of a parameter-free test, we might instance testing observations to see if they are distributed in a normal form, the mean and variance remaining unspecified. On the other hand, the typical goodness-of-fit χ^2, extensively discussed in Chapter 8 and defined by formula (27), is, at least in sufficiently large samples, completely independent of the form of the underlying distribution being tested and of the parameters in that distribution. Such tests are clearly distribution-free.

Nevertheless, it must be admitted that the terms *distribution-free* and *non-parametric* are widely used in the literature as interchangeable synonyms. We need not pursue any fine distinctions in the present discussion. The important point is to recognise when more robust methods of significance testing are called for, and to know some of the ways of dealing with the problem. Some of the more elementary applications are reviewed below.

15.2 PAIRED COMPARISONS

Let us reconsider, first of all, the *method of paired comparisons*, already discussed and illustrated in detail in section 6.2, with data as shown in Table 4. In the latter example we examined the relative effects of two analgesic drugs, A and B, on a small series of patients. Each patient was tested with both A and B, and the hours of relief obtained with each drug were recorded. Because of the great variation between patients, it was recommended that the relative advantage of one drug over the other, e.g. B over A, should be calculated for each individual patient, and that this series of relative advantages should be subjected to a t-test to see if there was a significant departure of the average relative advantage from zero. In the example we concluded that B was a significant improvement over A.

It was clear from Table 4 that not only were there big differences between patients, but such differences tended to persist whatever drug was used. Thus the observations for patient No. 6 were low for both drugs, and those for No. 7 were both high, although there was a relative advantage for B in each case. We avoided this correlation of measurements taken on any given patient by concentrating on the *relative advantage* for each patient. This automatically eliminated the systematic differences between patients, and allowed us to test the differences between the two drugs more directly. It also resulted in a much smaller residual variance s^2, because the variation between patients had been removed. This in turn meant that the test-statistic used was more sensitive to the particular comparison we were interested in.

Although this paired comparison test is not completely independent of the basic distributions of patient relief, it is clearly less sensitive to them than would be a straight comparison of the two means, μ_1 and μ_2, for the two series for drugs A and B taken separately. Not only have we eliminated the effect of systematic patient differences, but it is no longer necessary to assume that the hours of relief obtained are normally distributed. The t-test proposed will be exactly valid provided only that the relative advantages come from the same normal distribution, and this is a much weaker assumption. Moreover, whatever the distribution of the relative advantages, it can be shown theoretically that the t-statistic tends to normality with increasing sample size.

Although there are many theoretical complications and qualifications to the above statements, the net result in practice is that the paired comparison test described does have a good measure of robustness and can be used with a considerable degree of confidence in a wide range of applications. It is not, however, completely distribution-free. In the subsequent sections of this chapter we shall go on to discuss some tests that are genuinely independent of the initial assumptions about the basic frequency distributions of the observations.

15.3 SIGN TESTS

In the initial discussion in section 6.2 of the paired comparison test applied to two analgesic drugs we noted that Table 4 revealed that drug B did better for six patients, while drug A gave some improvement in two patients. We also mentioned in passing that there was some doubt as to whether this implied that we could assume B to be really better than A. The idea hinted at was that we might obtain some indication of the relative value of A and B simply by inspecting the *signs* of the differences shown in the right-hand column of Table 4. Let us now consider this concept more explicitly.

If the column of differences for $B - A$ showed eight positive values, and no negative ones, most people would rightly conclude, in spite of the smallness of the sample, that B was pretty consistently better than A. Would this be correct? And if so, why? It is intuitively obvious that, whatever the distribution of hours of relief obtained using analgesic drugs, positive and negative values would be expected to occur equally often on average if there was no difference between the drugs. In a small sample, therefore, we could ask whether a significantly high or low proportion of positive (or negative) values was in fact observed.

The occurrence of positive and negative values will in fact follow a binomial distribution with $p = \frac{1}{2}$, irrespective of the basic distributions of the primary observations. The test proposed will therefore be distribution-free. We thus have a situation which is identical to that discussed in Chapter 4 on the frequency of boys and girls in a family of given size, assuming the probability of a male or female to be exactly one-half. The chance of precisely eight positives would then be $(\frac{1}{2})^8 = 1/256$, and the chance of the opposite extreme of eight negatives would be the same. For a two-sided test to detect departures in either direction from the null hypothesis, we thus have $P = 1/128 = 0.78$ per cent. Such a result would be highly significant, confirming the intuitive interpretation.

However, suppose now that there were seven positives and one negative. It is not difficult to show (using formula (2) e.g. with $n = 8$ and $p = q = \frac{1}{2}$) that the total probability in the two tails is then $P = 18/256 = 7.0$ per cent. This result is no longer significant. The same conclusion applies even more strongly to the actual example of section 6.2 with six positives and two negatives. However, in the latter case we did obtain a clearly significant conclusion when using the t-test applied to the series of *quantitative* differences. This illustrates how the sign test is a less efficient test *when* the stronger assumptions are justified. But if we do not feel justified in making these stronger assumptions, we can fall back on the much more widely valid distribution-free sign test.

As usual, some judgment is required as to what method of analysis is appropriate and reliable in any given circumstances.

An additional point worth bearing in mind is that, since the sign test is easy to perform (using an exact analysis for small samples or the large-sample normal approximation to the binomial given in section 5.3 for bigger samples), it is often possible to obtain a rapid evaluation of a set of figures without detailed calculation. Thus, any paired comparison figures showing all positive or all negative deviations will be significant for a two-tailed test if there are six or more pairs of observations (the one-tailed version being significant for five or more). In such cases the significance is not in doubt, and elaborate calculations based on the t-test would add nothing further. Frequently, however, a more searching test is needed, and we can have recourse to the Wilcoxon test described in the following section. Of course, if in the event of significance we want to *estimate* the average relative advantage of the better drug, and perhaps set confidence limits, then we must carry out a more detailed quantitative analysis.

15.4 WILCOXON'S SIGNED RANK SUM TEST FOR ONE SAMPLE

Let us now carry the foregoing distribution-free approach a stage farther. As we have seen, a t-test analysis of the quantitative paired comparisons in Table 4 reveals a clearly significant difference between the two analgesics A and B. But a simple sign test, based on the occurrence of two negative and six positive differences, is nowhere near significant. This latter conclusion seems intuitively a little weak. Even if we are unwilling to accept the idea that a t-test is likely to be approximately valid, we might expect to obtain a somewhat stronger inference. This expectation is strengthened by close examination of the actual differences. Without making any very specific assumptions about the distribution of these differences, e.g. that they might be normally distributed, we can see that the two negative differences are both numerically very small, while some of the positive differences are quite large. There should therefore be a possibility of extracting more information from the sample, even in a distribution-free approach, to suggest that the distribution of differences is not symmetrical about zero but strongly skewed in the positive direction.

A very suitable method of investigation is provided by Wilcoxon's signed rank sum test, which operates as follows. First, put all the observations in ascending order of magnitude, ignoring the signs. Any zero values are ignored, and the remaining non-zero values are assigned the ranks 1 to n. If any of the observations are numerically equal they are each assigned an average rank calculated from the ranks that would otherwise have been used. Such equal ranks are said to be *tied*. We next calculate the sum T of the ranks of the positive observations. For moderate values of n we can refer directly to a table that shows, for appropriate significance levels, the extreme values of T which would need to be *attained* or *exceeded*. Appendix 6 give the exact values required up to $n = 25$.

Table 29 shows how the assignment of ranks actually works out for the differences appearing in the last column of Table 4. In these data we have eight primary observations, so that $n = 8$. These observations appear in the first column of the table in numerically ascending order of magnitude. The second column shows the relevant signs. The third column simply sets out the basic ranks from 1 to 8 in order. But we notice that there are two observations (actually of opposite sign) having the numerical value of 0.6. Since these tied values correspond to ranks 2 and 3, we must assign tied ranks of $2\frac{1}{2}$ to each of them. The adjusted set of ranks is therefore as shown in the fourth column of the table.

Table 29. *Assignment of ranks for applying Wilcoxon's test to data from Table* 4

	Numerical value of observation in ascending order of magnitude	Sign attached to the observation	Basic rank order	Rank adjusted for tied values
	0.5	−	1	1
Tied	0.6	−	2	$2\frac{1}{2}$
Tied	0.6	+	3	$2\frac{1}{2}$
	0.8	+	4	4
	1.2	+	5	5
	1.9	+	6	6
	2.3	+	7	7
	2.7	+	8	8

We easily calculate T as

$$T = 2\tfrac{1}{2} + 4 + 5 + 6 + 7 + 8 = 32\tfrac{1}{2}.$$

(If all the pluses and minuses had been exactly reversed we should have had essentially the same situation, but with the skewing in the opposite direction to give $T = 3\frac{1}{2}$.) Reference to Appendix 6 shows that the extreme values, corresponding to $n = 8$, that have to be attained or exceeded in either direction are 3 and 33 for significance at the 95 per cent level. Technically, therefore, the observed value $T = 32\frac{1}{2}$ does not quite achieve significance on a two-tailed test, though it is extremely close. This is a very much stronger result than that given by the simple sign test in section 15.3. If, of course, we had a prior reason for expecting analgesic B to be better than A, if anything, a one-tailed test would be appropriate. We should then enter the table in the 10 per cent column to find the corresponding 5 per cent level. Anything equal to, or greater than, 31 would be significant at this level.

When the number of comparisons, n, is too large for the table in Appendix 6 to be used, we can use a normal approximation. In effect, we take the normal variable with zero mean and unit standard deviation given by

$$d = \frac{|T - \frac{1}{4}n(n+1)| - \frac{1}{2}}{\sqrt{v}}, \tag{88}$$

where the numerator is the *absolute* value of the difference between T and its theoretical average, reduced by a 'continuity' correction of one-half, and v is the variance of the numerator.

If there are no ties we have

$$v = \frac{n(n+1)(2n+1)}{24}. \tag{89}$$

But when ties occur each group of t tied ranks reduces v by $(t^3 - t)/48$.

For the data in Table 29 we find

$$|T - \tfrac{1}{4}n(n+1)| - \tfrac{1}{2} = \left|32.5 - \frac{(8)(9)}{4}\right| - 0.5 = 14,$$

and

$$v = \frac{(8)(9)(17)}{24} - \frac{(2^3 - 2)}{48} = 51 - \frac{1}{8} = 50.875.$$

Hence, if we were attempting a normal approximation we should find

$$d = \frac{14}{\sqrt{50.875}} = \frac{14}{7.133} = 1.963.$$

This result is formally (see Appendix 1) just beyond the 5 per cent point, but of course a sample size of 8 is too small for reliable results to be obtained from a normal approximation. At the same time there is quite close agreement between the exact result obtained from the table in Appendix 6 and the large-sample approximation.

It should be emphasised that this Wilcoxon test essentially tests whether the random variable examined is distributed symmetrically about zero. A significant result suggests that the mean value is different from zero. The test can easily be extended to cover the hypothesis that a variable is symmetrically distributed about some value other than zero, μ say. In this case all we have to do is to subtract μ from each observation, and then use the test as before to see if there is a significant departure from symmetry about zero.

15.5 WILCOXON'S RANK SUM TEST FOR TWO SAMPLES

The ideas involved in Wilcoxon's distribution-free test for single samples can be extended to cover the comparison of two samples. Suppose we have two groups of observations given by the n values: $x_1, x_2, \ldots, x_n$; and the m values: $y_1, y_2, \ldots, y_m$. The null hypothesis is simply that the x's and y's have identical distributions, but we are interested in using a test that will tend to reveal whether the two distributions differ appreciably in location, i.e. whether they have different means.

All we have to do is combine the observations from both samples, put them in order of ascending magnitude and assign ranks. Tied values are given average ranks as in the Wilcoxon one-sample test described in the previous section. While doing this we must make sure that, although the samples are combined for ranking purposes, the x's and y's can be identified when we want. This is important, because the next step is to calculate the sum T of the ranks of all the x's.

We can now refer to published tables to find the values of T which would yield significance if they were *attained or exceeded in a more extreme direction*. Such tables are in principle used just as in the Wilcoxon single-sample test, but are more complicated in detail because of the two-way tabulation needed for dealing with possibly different sample sizes for the x's and y's. Tables are therefore not reproduced in this book but are easily available. A very useful source is in the *Scientific Tables* published by Ciba–Geigy, where the limits for T are shown for $n \leqslant 25$, $m \leqslant 50$ (note that in these tables $n = N_1$, $m = N_2$).

For sample sizes outside the scope of the tables there is a convenient normal approximation. We calculate the normal variable d, with zero mean and unit standard deviation, given by

$$d = \frac{T - \frac{1}{2}n(n + m + 1)}{\sqrt{v}}. \tag{90}$$

When there are no ties we have

$$v = \frac{1}{12} nm(n + m + 1). \tag{91}$$

When ties do occur we must use a modified formula for v which can be conveniently written as

$$v = \frac{nm(N^3 - N - R)}{12N(N - 1)}, \tag{92}$$

where

$$N = n + m, \tag{93}$$

and the reduction term R is found by adding together all the quantities $t^3 - t$ arising from each group of t tied values.

As an illustration of this method let us return to the data in Table 4 on the effects of the two analgesic drugs A and B. Let us suppose, for the sake of the example, that the observations could not be paired, and that we simply had two samples of data, the x's being shown in the second column of the table and the y's in the third column. In this case we have $n = m = 8$.

We first have to rank all sixteen observations, being careful, however, to maintain the distinction between the x's and y's. This step is shown in Table 30. We can easily calculate T as

$$T = 2 + 3 + 4 + 5 + 6 + 12 + 13 + 14 = 59.$$

(A useful check here is to distinguish between the sums of ranks given by T_1 for the x's and T_2 for the y's. We must then have $T_1 + T_2 = \frac{1}{2}N(N + 1)$. In the example $T_1 = 59$ and $T_2 = 77$, so that $T_1 + T_2 = 136 = \frac{1}{2}N(N + 1)$ if $N = 16$.)

Reference to the Ciba–Geigy tables with $N_1 = 8 = N_2$, gives the 5 per cent significance limits as 49 and 87. Since 59 is within these limits the result is not significant. This is hardly surprising here since we have ignored the information obviously present in the pairing of the observations from individual patients.

If we were to use the normal approximation in the situation with no ties indicated by formulae (90) and (91), we should have

Table 30. *Data on effects of two analgesic drugs treated as separate samples*

x-observation	y-observation	Rank
	1.0	1
1.2		2
1.6		3
2.8		4
2.9		5
3.2		6
	3.5	7
	3.6	8
	3.8	9
	4.8	10
	5.0	11
5.5		12
5.7		13
6.1		14
	7.3	15
	8.4	16

$$d = \frac{59 - (\frac{1}{2})(8)(8+8+1)}{\sqrt{\left[\left(\frac{1}{12}\right)(8)(8)(8+8+1)\right]}}$$

$$= \frac{59 - 68}{\sqrt{90.667}}$$

$$= -0.945,$$

a clearly non-significant result.

A final point to be mentioned in this section is that there are two other tests in common usage, bearing the names 'Mann–Whitney U-test' and 'Kendall's S-test'. Although there are certain differences of detail in presentation and calculation they are essentially equivalent to the Wilcoxon rank sum test described above. Further discussion is therefore not necessary here.

15.6 RANK CORRELATION

We have already discussed the general notion of correlation in Chapter 9, and looked in detail at methods of estimating correlation coefficients and carrying out tests of significance when the underlying distribution of the two variables involved could be assumed to be bivariate normal. The latter was seen to be a fairly severe restriction, which could sometimes be overcome by using the more widely applicable approach of regression analysis. However, such techniques still involve rather specific quantitative assumptions that may be quite unacceptable in many practical situations. Not only may assumptions based on normal distributions be invalid, but it often turns out that the observations themselves are of a very qualitative nature. Thus, in some cases it may not be possible to assign specific numerical measurements, though it may be feasible to marshal the observations into an ordered series and assign ranks, as in the foregoing sections.

This situation typically arises where subjective judgments are made with some degree of confidence, as with many investigations in the areas of psychology and education, and it is desired to have objective statistical tests of whether the associations alleged are likely to be more than merely chance occurrences.

If we want to think in terms of a coefficient of association that has at least some affinities with the correlation coefficient ρ of Chapter 9 certain characteristics are desirable. For example, the coefficient sought should lie between $+1$ and -1, taking the former value when there is in some sense a complete association between the qualitative variables observed and the latter value when the situation is reversed. We should also like the coefficient to be zero, at least in large samples, when either variable is distributed independently of the other.

To fix ideas suppose that we have a group of n individuals, each of which can be classified under each of two headings A and B. It is assumed that, so far as the A-classification is concerned for example, any individual measurements may have no objective meaning but it is possible to put the whole series of n measurements into a reliable order and assign ranks from 1 to n. A similar assumption is made about the B-classification.

We are thus faced with two rankings of the n individuals, and can ask to what extent these rankings are associated or correlated, and whether there is any significant departure from zero.

A suitable coefficient that is widely used is Kendall's rank correlation coefficient τ. This has the desirable properties mentioned above and can be defined as follows. The n individuals can be classified into $\frac{1}{2}n(n-1)$ distinct pairs. Let P be the number of pairs in which the two individuals are ranked in the same direction by both the A and B classifications, while Q is the number of pairs yielding rankings in opposite directions. (The illustrative example below will make the meaning of this definition quite clear.)

If we now put $S = P - Q$, Kendall's τ is given by

$$\tau = \frac{S}{\frac{1}{2}n(n-1)}. \tag{94}$$

A significance test for departures from the null hypothesis value of zero is most easily performed on S itself. For small values of n, e.g. $n \leqslant 10$, we can refer to the limits shown in Appendix 7, which gives the extreme values of S in either direction that must be *attained or exceeded* for significance to be achieved at the level indicated.

When n is beyond the scope of this simple table, we can use a normal approximation based on a normal variable d, with zero mean and unit standard deviation, given by

$$d = \frac{S}{\sigma_S}, \tag{95}$$

where the large sample variance of S is

$${\sigma_S}^2 = \frac{1}{18}n(n-1)(2n+5). \tag{96}$$

When employing this normal approximation for $n > 10$ it is as well to make a 'continuity' correction by replacing S by $S - 1$ if S is positive, and by $S + 1$ if S is negative.

The possible existence of tied rankings is an awkward complication in this test. When $n \leqslant 10$ tables exist for any number of tied pairs or tied triplets (Sillitto, G. P. (1947), *Biometrika*, **34**,

36), but the tabulations would have to be very extensive to cover larger runs of tied values. When $n > 10$ we can use the normal approximation described above but with a modification of the variance $\sigma_S{}^2$. The formula is somewhat complicated, and when required perhaps it would be best either to seek help from a professional biostatistician or refer directly to the book by Kendall & Gibbons, *Rank Correlation Methods*.

As an example in the biological field let us suppose that we wish to test whether there is any correlation (in a general sense) between the adult weights of fathers and sons in some animal where weight is extremely variable. Consider, therefore, the eight pairs of values as shown in Table 31, where x represents the father's weight and y the son's weight in each pair. There does seem to be an obvious correspondence between the x's and y's, but the occurrence of three low values with small variations (Nos. 1, 2 and 3) and five large values with much larger variation (Nos. 4 to 8) precludes any possibility of using the correlation approach of Chapter 9, or the regression approach of Chapter 10. Bivariate normality or equal variances about a regression line are clearly not acceptable assumptions. Some improvement might be achieved by using transformed variables (as indicated at the end of section 10.1), but a quick distribution-free test can be made quite easily using Kendall's τ.

We first rank the quantitative x-observations and obtain the order shown in the fourth column of Table 31. The rankings for the y-observations are similarly shown in column 5. It can easily be seen how the first three pairs fall into one group of rankings, and the last five pairs into another. The same conclusion can be seen equally clearly by plotting the (x, y) pairs on a graph.

Next, we rearrange the pairs in two columns so that one set of ranking, e.g. for the x-observation, appears in the standard ascending order of magnitude 1, 2, 3, 4, etc. This rearrangement is shown in columns 6 and 7 of the table.

The two quantities P and Q, needed to calculate $S = P - Q$, are now easily found as follows. P is found by taking each ranking of column seven in turn and counting how many rankings below have a higher value. Thus

$$P = 7 + 5 + 5 + 3 + 1 + 2 + 0 + 0 = 23.$$

Table 31. *Correlation of animal weights in father–son pairs*

Pair number	Actual weight		Ranking		x-ranking in natural order	
	Father (x)	Son (y)	x	y	x	y
1	1.0	1.2	1	1	1	1
2	2.2	2.6	3	2	2	3
3	1.5	3.1	2	3	3	2
4	9.4	10.3	7	8	4	5
5	7.3	8.5	5	7	5	7
6	8.1	3.4	6	4	6	4
7	10.0	7.2	8	6	7	8
8	6.5	6.7	4	5	8	6

The value of Q is obtained similarly by counting rankings with lower values. Thus

$$Q = 0 + 1 + 0 + 1 + 2 + 0 + 1 + 0 = 5.$$

As a check, we may use the fact that

$$P + Q = \tfrac{1}{2}n(n - 1).$$

In this example $P + Q = 28 = (\frac{1}{2})(8)(7)$, as required. For a significance test we note that

$$S = P - Q = 18.$$

Reference to Appendix 7 shows that for $n = 8$ the result is significant at the 5 per cent level. (Note that the significance level P, i.e. the P value, is not to be confused with P in the definition S.)

As an illustration of the normal approximation method, we should first make the continuity correction to S reducing the positive value of 18 by one unit, i.e. to $S = 17$, and then calculate the variance from formula (96). We easily find

$$\sigma_S{}^2 = \left(\frac{1}{18}\right)(8)(7)(21) = 65.333.$$

Hence

$$\sigma_S = 8.083,$$

and from formula (95)

$$d = \frac{S}{\sigma_S} = \frac{17}{8.083} = 2.10.$$

This is comfortably outside the 5 per cent point for a two-tailed test, though $n = 8$ is rather small for the normal approximation to apply very accurately. We thus have the same conclusion as in the exact test above.

It should finally be mentioned in passing that there is also another, older, measure of rank correlation, namely Spearman's rank correlation coefficient. This also provides distribution-free tests of association and is not very different from Kendall's τ in general consequences. The latter does, however, have certain practical and theoretical advantages.

15.7 CONCLUDING DISCUSSION

It can be seen from the foregoing exposition that non-parametric and distribution-free tests offer, in appropriate circumstances, very effective means of avoiding unnecessarily restrictive assumptions that insist, for example, on the normality of an underlying distribution, or, even worse, the existence of bivariate normality. We might expect to have to pay a price for the increased generality and validity of distribution-free tests. But, surprisingly enough, this is not always the case.

Take, for instance, Wilcoxon's rank sum test for comparing two samples (see section 15.5). If the underlying distributions *are* normal with the same variances, the best test to use for comparing the means is the standard t-test (as recommended in section 6.3). But the Wilcoxon test is still 96 per cent efficient. (We define 'efficiency' here in terms of the ratio of sample sizes needed to detect a given small difference in the location of the distributions.) Moreover, if the underlying distributions are not normal, the efficiency of the Wilcoxon test never falls below 86 per cent and for some distributions that are very different from normal the relative efficiency may be much greater than 100 per cent. This means that in the latter case the Wilcoxon test achieves the same result as the t-test with a smaller number of observations.

Many people have taken such arguments to imply that, in the comparison of two samples, the Wilcoxon test is always preferable. It must be remembered, however, that all of these distribution-free methods are only *tests*. Important though these are, we must not lose sight of the further need for *estimating* specific parameters, together with their standard errors, and interpreting the results. This entails assumptions about the distributions involved. Moreover, the advantages of a distribution-free approach, while often appreciable, are not always as clear-cut as in the Wilcoxon two-sample test.

Generally speaking, it is probably safest to apply distribution-free methods when there is some doubt about the approximate validity of the usual assumptions underlying standard tests. Sensibly used in this way such methods can greatly enhance our capabilities of extracting significant information from data whose interpretation might otherwise remain in doubt.

16

Notes on numerical calculation, calculators and computers

16.1 INTRODUCTION

It is proposed in this chapter to give a short account of some of the special points worth noting in connection with the fundamental business of carrying out basic statistical calculations, with regard to both the arithmetical aspects and the equipment required. There are many important details which can be easily overlooked by those who are mainly concerned in dealing with the more elementary work themselves, and who do not need to become involved in the niceties of advanced computation. Unless calculations are sufficiently accurate it is obvious that the statistical tests and analyses based on them may be completely vitiated. Nevertheless, one wants to avoid laborious complexities as far as possible. We shall therefore consider below some of the most important features to which special attention should be given, in particular, avoiding mistakes, achieving sufficient numerical accuracy, exhibiting the precision of statistical results, and choosing the most suitable level of automatic assistance, whether this be a simple or advanced pocket calculator, or a computer with specialised statistical software.

16.2 HOW TO AVOID MISTAKES

Before going on to look at computational details, we shall first consider some aspects of the general question of obtaining accurate calculations. It is very easy, even for those who are experienced in numerical work, to make small slips in reading or writing figures on a worksheet, in transferring numbers to or from a desk calculating machine or pocket calculator, in attaching plus and minus signs, or in interpreting brackets in algebraic formulae, etc. Naturally, these errors decrease with practice, but the possibility of mistakes in a long sequence of calculations is always fairly high. Although one wants calculations to be as free from error as possible, failure to discover a mistake is much worse than merely making the mistake in the first place. It is therefore essential to operate some kind of check on the accuracy.

The best kinds of check are those which are, so to speak, built in to the series of calculations. Perhaps the simplest example is when we have data in the form of a contingency table. Adding the entries by rows gives the row totals. Similarly, adding by columns gives column totals. The grand total is the sum of the row totals or the sum of the column totals. So both of these alternative calculations must give the same answer if correctly performed. Again, in a factorial experiment it is often convenient to find the treatment sum of squares in two ways: first, in the standard way from treatment totals; second, by adding the individual factorial contributions from main effects and interactions.

If a cross-check with the same numerical quantity being calculated by two different methods is not available, we must fall back on simply repeating the calculations. This is not a very satisfactory procedure if the repetition is performed by the same individual, although some people do seem to find that it usually provides them with a sufficiently reliable check. The main objection is that one is very prone to repeat one's errors. This difficulty can be off-set to some extent by doing the various additions, multiplications, etc., in a different order. It is also advantageous to let as much time as possible elapse before the repetition of calculations, in order to reduce the chance of making the same mistake twice.

Much the best kind of repetition is when two different people independently perform the whole of a calculation, only comparing answers afterwards. It is much less satisfactory for a second individual to go through a calculation already performed by somebody else. There seems in this case to be a considerable unconscious bias leading the checker to misread figures at each stage in his or her own calculations so as to make them agree with the original version if this is different.

In practice one often has to be satisfied with something less than the ideal. For example, except where computing staff are available, it is usually not possible to have calculations performed independently by more than one individual. However, it is important to pay special attention to the elimination of any mistakes that may have crept into one's numerical work.

It is also important to realise that mistakes are very rarely made by calculators and computers: practically all errors are due to purely human failings – typically feeding in misread data, pressing wrong keys, using unsuitable software, misjudging outputs, etc. In addition, the use of a computer offers extended opportunities for storing vast amounts of data. Great care must therefore be exercised to ensure that any database is accurately recorded.

16.3 RELATIVE NUMERICAL ACCURACY

In the last section we have considered the question of avoiding or detecting specific mistakes such as misreading figures, omitting minus signs, misinterpreting expressions in brackets, and so on. These are gross errors which can be eliminated by exercising greater care and making proper checks. There remains, however, the difficulty of the relative numerical accuracy of the work, quite apart from straightforward blunders. Just as measurements may be recorded only to the nearest gramme or centimetre, so subsequent calculations may retain at each stage only so many decimal places or significant figures. These distinctions are discussed in more detail below. The point here is that small errors of approximation are continually introduced into calculations, and may, if we are not careful, accumulate to a dangerous extent.

Let us first take the number of *decimal places* in any quantity. This is, of course, simply the number of digits retained to the right of the decimal point. Thus 6.32 has two decimal places. This number may be a direct measurement, or it may be the end product of some calculation. Measurements usually have some lower limit of accuracy: we measure to the nearest metre or hundredth of a metre, etc. If we can reckon on taking certain readings to the nearest hundredth of a metre, then 6.32 m means that the 'true' measurement is somewhere between 6.315 and 6.325, since values slightly less than or slightly greater than these two quantities, respectively, would have been recorded as 6.31 or 6.33. There is therefore a maximum error at this stage of 0.005 m, a relative value of about 1 in 1200. The same principle applies when any number with several figures in it is 'rounded off' to a smaller number of figures. Incidentally, it is nearly always preferable to round off to the *nearest* adjacent value, and not 'round up' or 'round down' systematically. Thus 6.32 should be rounded off to 6.3 (not 6.4) if we wanted to retain only one decimal place. Numbers like 6.35 which are exactly half-way between the alternatives of 6.3 and 6.4 are a little more trouble. If we round such numbers systematically up or down, there is always the danger of a small bias creeping in. It is much better to round numbers ending in 5 up and down at random, so far as possible. In practice it is usually sufficient to round up numbers ending in 15, 35, 55, 75 and 95, and to round down numbers ending in 05, 25, 45, 65 and 85 (or vice versa). If several quantities of this type occur in a series, it is probably quite safe in general to round up and down alternately.

The next matter to be mentioned is that of *significant figures*. Broadly speaking, the number of significant figures is the number of figures involved, not counting zeros used to 'fill in' before or after the decimal point. Thus, 6.32 has three significant figures; so have 0.00632 and 632 000. The only exception is when a number like 632 000 is really accurate down to the units place. If the weight of an animal is measured as 1000.2 grammes and we decide to record this to the nearest gramme, then we shall write

1000 grammes; this figure will actually be accurate to *four* significant figures and not to only one. Whenever any confusion is likely to arise, a special note or comment should be made as a reminder. It is obvious that high relative accuracy is related to the number of significant figures rather than to the number of decimal places. Thus 0.00001 might be anything from 0.000005 to 0.000015, and it is the one significant figure that gives rise to this low relative accuracy in spite of the five decimal places. The relative accuracy of 0.00001 is, therefore, at worst 50 per cent. But with three significant figures, say 0.0000102, the relative accuracy is at worst only $\frac{1}{2}$ per cent.

When performing a series of calculations we want to end up with numbers such as means, standard deviations, t, χ^2, etc. that are sufficiently accurate for our purpose. If we want a χ^2 to be accurate to two decimal places (this is usually about right for making comparisons with tables of percentage points), we must arrange the appropriate calculations so as to achieve adequate accuracy at each stage. Suppose we are deriving a homogeneity χ^2 made up of several components. If each component is accurately computed to four decimal places, then in most cases the total will be accurate to at least two and can be safely rounded off to this extent.

Again, if we want to quote regression coefficients accurately to about, say, 1 per cent, the calculations leading up to them must be at least as accurate as this, and preferably more so. It is difficult, if not impossible, to lay down general rules to cover all cases since the accuracy achieved often depends on the actual numbers involved. But, with the important exception described below, it is usually adequate to retain two *more* significant figures at each stage than we aim at in the final answer.

The exception just mentioned is when we find the difference between two numbers that are approximately equal. A typical occurrence is in the formation of the sum of squares about the mean. Suppose the crude sum of squares is 20 186.2 and the corresponding correction factor is 20 152.1. Both of these numbers contain six significant figures, but the difference, i.e. the sum of squares about the mean, is 34.1 and has only three significant figures. This is hardly good enough. We probably ought to retain at least two more figures in both the sum of squares and the correction factor to obtain sufficient accuracy in the difference. There is no difficulty with most pocket calculators and computers in carrying these extra digits. In the present example there would have been a considerable saving of effort if we had started off with a working origin somewhere near the mean. All the observations would then have been much smaller, and the required accuracy for the sum of squares about the mean could have been obtained with less trouble.

These troubles will often be avoided if one uses a pocket calculator which will automatically present the mean and standard deviation of a series of numbers that have been appropriately keyed in (as referred to in section 16.5 below). However, the same possibility for error described above still exists, if we have, for example, a series of large numbers all rather close together. Some calculators operate with only eight digit accuracy, though there are pocket calculators with as many as thirteen or more. Thought must therefore be given to ensuring that the results obtained, especially standard deviations, will have a sufficient number of *significant figures* to give the accuracy required. As before, a good solution is to choose a working origin somewhere near the mean value.

16.4 THE PRECISION OF RESULTS AND STANDARD ERRORS

As pointed out above, it is as well to carry out the main calculations of an analysis to a substantial degree of accuracy, even if we require less precision in the final answer; one can always round off the answer to fewer significant figures if necessary, but it is impossible to obtain extra ones without fresh calculation. The question now arises as to the relative accuracy to which means, standard deviations, regression coefficients, etc., should be quoted. The answer depends partly on the use to which they are to be put and partly on the statistical reliability, which is limited by the sampling variation involved.

To begin with we should normally use a sufficient number of

significant figures to distinguish the sort of differences that seem important to us. Basic measurements are often considerably affected by experimental error. It is difficult, for example, to record the heights of human beings to a greater accuracy than the nearest millimetre, and some observers would not think it worth while attempting more than the nearest 10 millimetres. This means that basic readings are numbers such as 1.67, with three significant figures at the most. (Admittedly, if we work in half centimetres we need four figures for some measurements, e.g. 1.675, but the fourth does not really reflect the true accuracy.)

In calculating means and standard deviations, we shall use perhaps two or three extra significant figures so as to ensure that the purely numerical work introduces no appreciable error in addition to the original experimental error and the sampling variation involved. In the process of calculating mean values, both the rounding off errors of measurement and the sampling variation affecting individual values tend to cancel out. The relative accuracy of the means is, therefore, usually much higher than that of single observations. We should be quite justified, in general, in quoting average values to at least an additional figure in samples of moderate size, e.g. 4.68 in the example of section 3.4.

Such results should also be considered in relation to the relevant standard error. In section 3.4 we had a mean of 4.68 ± 0.11 cells. This form is about right for most practical purposes. It is not usually worth giving the standard error to more than about two significant figures. The reason for this is that any additional numerical accuracy would be negligible compared with the degree of statistical variation represented by the standard error itself. An alternative rule is to retain no more significant figures than are necessary to keep the rounding-off errors less than one-tenth of the standard error. Having fixed the number of significant figures in the standard error, we should clearly write the associated mean to the same number of decimal places. Suppose in the example we had calculated the mean as 4.683 and the standard error as 0.114, then we can contract the latter to 0.11 and then write the former to two decimal places as well, giving 4.68 ± 0.11. If it should happen that the mean had insufficient places to be rounded off in this way, then we should have to recalculate to a greater accuracy.

Considerations similar to the above can be applied equally well to other statistical indices such as correlation coefficients, regression coefficients, etc. All the time one has to strike a balance between overburdening one's work with unnecessary extra digits and having too few figures to ensure sufficiently accurate results. If one is compiling tables of means for general discussion, the number of significant figures should be reduced to the minimum required for intelligent interpretation. But if further mathematical work is to be based on published results, then, in order to be on the safe side, it is safer to include slightly more significant figures than would otherwise be used.

16.5 CHOOSING CALCULATORS AND COMPUTERS

As indicated in the introduction, 'The succession of subjects introduced is intended to provide a graded course of instruction in elementary statistical methods'. This means that we have been primarily concerned with statistical principles and their application to a variety of relatively simple problems. In addition, mathematical symbolism has been reduced to a minimum and used only in the role of short-hand instructions designed to facilitate calculation. Many worked examples have been provided, and indicate not only the underlying logical and arithmetical structure but also the means of actually calculating the required statistical results. In many cases the latter can be achieved employing only elementary arithmetic. But, as frequently mentioned in the text, the use of a good pocket-sized electronic calculator is highly desirable both for speed and accuracy, as well as efficiency. For more complicated applications, such as multiple regression, something more sophisticated is indicated – either an advanced, possibly programmable, pocket calculator, or some kind of computer with specialised statistical software, particularly when large data sets are involved. This is a very big subject and needs to be reviewed in depth, but in the present context we can at least look at the principles of choosing a suitable level of automatic computation to start with and consider how to develop more elaborate undertakings in the future.

Nearly everyone nowadays is familiar with the simplest pocket calculators that will carry out automatically the basic arithmetical operations of addition, subtraction, multiplication and division. But for the simplest statistical applications rather more than this is required. Thus, for the whole range of examples presented in this book, we need automatic calculation, using one or two keys at most, for each of the functions $\sqrt{x}$, $1/x$, x^2, $x!$ and e^x, where x is any number likely to occur in the appropriate context. It is also desirable to have the inverse of the exponential function e^x, namely $\ln(x)$, where ln indicates the natural logarithm $\log_e$. In some numerical work it can be more convenient to work with logs to base 10, i.e. $\log_{10} x$, with inverse 10^x. And for some applications the power function y^x can be helpful. But the last three functions cited are less important than the previous ones. Again, in certain kinds of scientific work and some statistical techniques trigonometrical functions are intrinsically important, in which case we need single keys for sin, cos and tan, as well as their inverses $\sin^{-1}$, $\cos^{-1}$, $\tan^{-1}$, although none of these functions has appeared anywhere in this book.

So far we have been considering the technical level required of a pocket calculator in carrying out the basic *arithmetic* (with one or two exceptions mentioned below) for the elementary statistical methods presented. But more is really needed for most easy statistical calculations. Even the simple case of means and variances, or standard deviations, can become extremely laborious and prone to error when a long series of numerical observations is involved. It is important therefore to ensure that your calculator has the additional facility of accepting a long string of numbers and then providing immediate displays of mean value and standard deviation merely by pressing the relevant keys. Many calculators will also accept a series of number pairs, e.g. (x_i, y_i) for $i = 1, \ldots, n$, and then automatically supply the estimated correlation coefficient r; or the fitted regression line $y = a + bx$, along with the concomitant means and variances (including residual variance about the regression line), thus providing everything for a variety of significance tests.

This is about as far as one can go without getting into deeper mathematical complexities, or avoiding these by taking advantage of programming approaches used in all computers and some advanced pocket calculators. A borderline case, referred to above in Chapter 14, is the common situation of multiple regression with two or more independent variables. Indeed, with only two independent variables, as in section 14.3, explicit formulae are available for parameter estimates, standard errors and residual variance. Calculation is therefore fairly straightforward, provided that our calculator will readily accumulate all the necessary sums of squares and products about mean values. But with three or more independent variables, as in section 14.4, we need the additional facility of solving the simultaneous equations in formula (83). If the calculator has the built-in ability to solve such equations, and/or built-in matrix algebra including matrix inversion, then the procedure is not difficult provided that the elementary matrix operations required are properly understood at least in principle.

Logically, the next stage to be considered is the use of programmable pocket calculators. Some of these, e.g. those produced by Hewlett Packard, Texas Instruments, Casio, etc., are extremely powerful in their own right, as well as being able to interface with more sophisticated equipment. In addition to the operations already mentioned above we can now, with such machines, solve nonlinear equations, carry out numerical integrations involving advanced mathematical functions, etc. *Ad hoc* programming of extensive mathematical investigations is possible, as is having access to an ever-increasing supply of ready-made programs from manufacturers and user groups. Software packages on magnetic cards are available for some models, and in many cases cards can be used to increase the calculator's storage capacity. Generally speaking, advanced programmable calculators are of most use to investigators with intricate mathematical problems either of a purely theoretical nature or arising in the context of relatively limited data sets. For example, the maximum-likelihood estimation of several parameters in a complex dynamic model, with subsequent validation, simulation and forecasting, can become a comparatively straightforward undertaking.

Those who are not primarily concerned with mathematical technicalities, but are, or have become, familiar with the elementary statistical methods described in this book, may wish to continue with the practical application of more complex methods which could be understood in principle without the necessity of delving into underlying intricacies. This is the point at which some form of computer comes into its own.

Let me say at once that this does *not* involve a big intellectual jump forward (though some may feel a psychological barrier). On the contrary, it makes a lot of things much easier. But it is essential to operate with levels of hardware and software that are appropriate to one's interests and abilities. A full discussion of these matters would require at least a whole book (of which there are many available). So in the present context let us look at just the main features.

If you are working or studying in a university department or some other research institution the actual basic equipment or hardware may well be determined by general administrative considerations. But, whatever is provided in terms of stand-alone machines or terminals linked to a large central mainframe, or a combination of both arrangements, your chief concern will be with the availability of appropriate statistical software to enable you to undertake a wide range of statistical investigations: data management and exploration, alternative forms of data analysis, multiple regression with many independent variables, experimental designs and factorial arrangements to fit widely varying circumstances, sophisticated forms of the analysis of variance, maximum-likelihood estimation, high-grade computer graphics, etc. Moreover, you will expect to be able to handle and store large data sets far beyond the capabilities of even the most advanced programmable pocket calculator. With the right equipment you will find it easy to access other people's data, either by acquiring a version already stored on diskette or by direct entry to a stored database if you are connected on-line.

A key question is, of course: 'How easy is it to achieve all this?' The short answer is: very easy if you start with some well-tried software that is not over-ambitious, but is *user-friendly*. Until fairly recently users were liable to be overwhelmed by having to study in depth reference manuals often running into hundreds of pages of obscure technical prose. Although it is important to have a good clearly written guide for occasional reference, a user-friendly software package provides an immediate interaction between user and computer. The user is helped to achieve his or her objectives by a series of multi-choice questions presented on the visual display. These deal with such matters as type of data available, how to enter it, kind of statistical analysis to be used, form of output required, etc.

By now there are hundreds of statistical software packages available: new ones are constantly appearing and older ones are regularly improved and updated. Ultimately, a personal choice has to be made, and it is almost invidious to mention specific package names. However, as a starting point it can safely be said that MINITAB is widely used, is regarded as very user-friendly, and will run on a wide range of different types and levels of computer equipment. There is also a specific (and very much cheaper) student-oriented version. Another package that can also be obtained at student level is KWIKSTAT, but there are many others. At a more sophisticated level there is a great variety of packages to choose from. One widely used is SPSS, and anyone wanting a relatively easy introduction could start with the cheaper student version. Other well-known names are SAS, GLIM, BMDP, etc.

However, whether you are working at student level in elementary applied statistics or are actively engaged in biological research, the best advice about embarking on computer-based statistical investigations is obtained by consulting friends and colleagues who are professional biostatisticians with considerable practical experience of biological applications and the use of different kinds of statistical software.

If you are thinking of acquiring for private personal use a pocket calculator, or more seriously a small computer, for some of the applications described in this book, it should be appreciated that considerable care should be exercised, especially if you intend to go on to more advanced work. Even small computers, such as personal computers (PCs) or the more convenient and even smaller lap-tops, which can be just as powerful, are liable to be relatively expensive – as can really efficient software. Moreover, you have to be sure that the software you want to use will run properly and efficiently on the hardware you have in mind. Again, it is essential to check this out with knowledgeable friends and colleagues before making a decision. In short, some form of bench-testing must be carried out.

Finally, let me repeat the strictures given at the end of section 16.2, namely, that calculators and computers very seldom make hardware mistakes. Practically all errors are introduced by human operators misreading their data, hitting the wrong keys, choosing inappropriate software, misunderstanding outputs, etc., not to mention errors of scientific judgement such as false or inadequate assumptions, faulty logic and reasoning, failure to validate models, etc. Very occasionally, it is true, there can be errors in software obtained from commercial sources, but these are rare and can be largely avoided by sticking to widely used packages produced by well-known and highly reputable companies.

Suggestions for more advanced reading

Tables

CIBA–GEIGY LTD. (1982). *Scientific Tables Volume 2: Introduction to Statistics, Statistical Tables, Mathematical Formulae* Basle: CIBA–GEIGY Corporation.
FISHER, R. A. and YATES, F. (1963). *Statistical Tables for Biological, Agricultural and Medical Research* Edinburgh: Oliver & Boyd.
LINDLEY, D. V. and SCOTT, W. F. (1984). *New Cambridge Elementary Statistical Tables*. Cambridge: Cambridge University Press.
MURDOCH, J. and BARNES, J. A. (1986). *Statistical Tables for Science, Engineering, Management and Business Studies*. Macmillan.
PEARSON, E. S. and HARTLEY, H. O. (1966). *Biometrika Tables for Statisticians, Vol. 1*. Cambridge: Cambridge University Press.

Statistical Methods

ARMITAGE, P. and BERRY, G. (1994). *Statistical Methods in Medical Research* (Third edition). Oxford: Blackwell Scientific Publications
COCHRAN, W. G. (1977). *Sampling Techniques* (Third edition). London: Wiley.
COCHRAN, W. G. and COX, G. M. (1957). *Experimental Design*. New York: Wiley.
FINNEY, D. J. (1980). *Statistics for Biologists*. London: Chapman & Hall.
FISHER, R. A. (1990). *Statistical Methods, Experimental Design, and Scientific Inference*. Oxford: Oxford University Press.
KALTON, G. (1983). *Introduction to Survey Sampling*. London: Sage Publications.
KENDALL, M. G. and GIBBONS, J. D. (1990). *Rank Correlation Methods*. (Fifth edition). London: Edward Arnold.
MEAD, R. (1990). *The Design of Experiments*. Cambridge: Cambridge University Press.
SOKAL, R. R. and ROHLF, F. J. (1994). *Biometry*. New York: W. H. Freeman.
YATES, F. (1987). *Sampling Methods for Censuses and Surveys* (Fourth edition). London: Edward Arnold.

Summary of statistical formulae

This summary is intended for use as a quick reference guide by the reader who already has some knowledge of statistics. It cannot be emphasised too strongly that standard formulae should not be applied blindly without some understanding of their suitability. Nevertheless, many workers who have already acquired a basic training in statistics, either from reading the main text of this book or elsewhere, will frequently require only to have their memories refreshed. This is the purpose of the summary. As an aid to practical applications, a short guide is included which leads automatically to the section of the summary actually required. A list of the chief symbols, used here and elsewhere in the book, is also provided. In most cases there are references to the extended discussions and worked examples appearing in the body of the book. The formulae are numbered so as to correspond with those given in the main text. The sections of the summary are numbered in heavy type, e.g. **7**, so that references to these are easily distinguished from references to sections in the main text or to formulae. If there is any doubt as to the application or interpretation of any formula, the reader should refer back to the full discussion and also bear in mind advice given on the use of pocket calculators or computers with special statistical software.

LIST OF CHIEF SYMBOLS USED

a, b, c, d	used for observed frequencies in contingency tables, etc.
b	estimated regression coefficient; number of blocks in randomised block design.
b_1, b_2, etc.	estimated partial regression coefficients.
B_j	sum of all yields in jth block of randomised block design.
β	true or population regression coefficient.
β_1, β_2, etc.	population partial regression coefficients.
c	estimated covariance; number of columns in contingency table.
c_{00}, c_{0j}, c_{ij}	sums of squares and products about means in partial regression.
C	correction factor for sums of squares in analysis of variance.
χ^2	chi-squared.
d	normal variable with zero mean and unit standard deviation.
e	base of natural logarithms.
exp	exponential, used in normal curve (1) and Poisson distribution (3), and bivariate normal distribution (34).
E	expected frequency in any cell of a contingency table.
f, f_i	numbers of degrees of freedom.
F	variance ratio.
G	grand total of all yields in experimental designs.
h	width of grouping interval in Sheppard's correction (8).
k	observed fraction in a binomial sample; number of varieties in completely randomised design.
m	population mean of Poisson distribution.
M, M_B, M_T	mean squares for varieties, blocks and treatments in analysis of variance.

μ, μ_x, etc.	true or population means of distributions.
n, n_i, N	numbers of observations in samples.
O	observed frequency in any cell of a contingency table.
$p = 1 - q$	expected fraction in a binomial sample.
P	significance level actually achieved by data.
r	estimated correlation coefficient; number of rows in contingency table; number of replicates in factorial design.
r_{12}, r_{13}, etc.	estimated correlation coefficients.
$r_{12.3}$, etc.	estimated partial correlation coefficients.
ρ	true or population correlation coefficient.
$\rho_{12.3}$, etc.	population partial correlation coefficients.
s, s_1, s_x, etc.	estimated standard deviations.
s', s'_1, s'_x, etc.	estimated standard deviations corrected for grouping (Sheppard).
s^2, s_1^2, s_x^2, etc.	estimated variances.
s'^2, s'^2_1, s'^2_x, etc.	estimated variances corrected for grouping (Sheppard).
σ	true or population standard deviation of a distribution.
σ^2	population variance of a distribution.
$\sum$, $\sum_1$, $\sum_2$	summation symbols.
t	'Student's' t; number of treatments in randomised block design.
T_i	sum of all yields for ith treatment in randomised block design.
τ	Kendall's rank correlation coefficient.
v	variance
x, x_i	observed measurements; independent variables in regression.
$\bar{x}$	mean of a sample of measurements x.
y	observed measurements; dependent variable in regression.
$\bar{y}$	mean of a sample of measurements y.
z, z_1, z_2, etc.	transformed values of observed correlation coefficients.
ζ, ζ_1, ζ_2, etc.	transformed values of population correlation coefficients.

GUIDE TO SUMMARY

Types of distribution

Normal (or Gaussian): This frequently arises, at least approximately, when we have a continuous measurement such as height or weight, for which the distribution is symmetrical and clusters about some central, average value. Observations become increasingly rare as we move farther from the average. (See section 2.1.)

Binomial: This occurs when each individual in a sample can be recognised as having, or not having, a certain character. We may consider, for example, having the proportion or percentage of male animals in a given group. (See section 2.2.)

Poisson: This often results when a particular kind of event is rather uncommon, but a sufficient number of instances are examined for several cases to be observed. Typical examples are the number of cells in a haemocytometer square, the number of bacterial colonies on a plate and the number of radioactive atoms disintegrating in a certain volume during a given time. (See section 2.3.)

Note on one- and two-tailed tests (See sections 5.4, 6.5.)

Tables of the normal distribution (often represented by d in this book) and of t usually give percentage points covering both tails of the distribution. To use these tables for *one-tailed* tests, we must enter the tables with *twice* the probability level actually required, i.e. 10 per cent in tables for a true level of 5 per cent.

On the other hand, tables of χ^2 and the variance-ratio F are usually presented in one-tailed form. To carry out a two-tailed test we must enter the tables at *half* the required probability, i.e. $2\frac{1}{2}$ per cent in tables for a true level of 5 per cent.

How to find the type of analysis required

First examine the data and decide which of the four following main types of problem is involved:

A. Problems concerned with the examination of a single sample of measurements, including comparison with a standard, or comparisons between two or more samples of measurements, especially with regard to averages or percentages.

B. Problems involving the classification of data into groups: comparing observed frequencies with theoretical ones (goodness-of-fit tests); contingency tables showing a two-way classification into groups; tests of homogeneity between several samples, especially when data can be exhibited in contingency tables. (χ^2 is the chief tool here.)

C. Problems dealing with the association of measurements, e.g. the *correlation* of pairs of variables and *partial correlation* of sets of more than two variables; also *regression* and *multiple regression* showing how changes in one particular variable are related to changes in one or more other variables.

D. Problems involving particular kinds of experimental designs: comparisons between the means of several groups of measurements; experiments containing blocks of units specially chosen to eliminate the effect of large variations in the experimental material; factorial experiments designed to test many factors simultaneously.

Procedure for Type A

For problems of Type A first judge what kind of basic distribution is likely to be involved, i.e. whether it is normal, binomial or Poisson. In some cases, the observations may be continuous, but clearly non-normal, or even rather qualitative in nature, suggesting a non-parametric or distribution-free approach. Secondly, decide whether there is only a single sample to be examined, and whether it is to be compared with a standard or whether there are two or more samples to be compared. Thirdly, determine whether the samples are large or small. Then refer to Table 32 in order to find the relevant section of the summary, shown in heavy type.

Procedure for Type B (χ^2 tests)

For general goodness-of-fit tests, comparing observed frequencies in groups with theoretical values: see **15**.

Table 32.

Type of distribution	Size of sample	One sample: comparison with a standard	Comparing two samples	Comparing several samples
Normal				
1, 2	Large	**3, 4**	**5**	**35**
	Small	**6**	**7**	**35**
Binomial				
8	Large	**9, 10**	**11**	**20, 17**
	Small	(See *Biometrika Tables*)	**20, 18, 19**	**20, 17**
Poisson				
12	$n\bar{x}$ large	**13**	**14**	**16**
	Single small values	(See *Biometrika Tables*)	(See *Biometrika Tables*)	**21**
Non-normal or unknown				
	Large or small	**38, 39**	**40** (See Ciba–Geigy, *Scientific Tables*)	–

For a two-way classification involving r rows and c columns: see **16**. Special case with only two rows: see **17**. Special case of 2×2 tables: see **18**; or if numbers are small: use **19**.

Homogeneity tests on several samples, each divided into a number of groups, may often be carried out by exhibiting the data in the form of contingency tables, e.g. taking samples as a row-classification and groups as a column-classification (or *vice-versa*). We then analyse as above, i.e. as in **16** to **19**. For binomial samples: see **20** and **17**; for Poisson samples: see **16** for large numbers, or **21** for single small values.

In combining the results of several χ^2 tests, the additive property of χ^2 and the normal approximation may be useful: see **22**.

Procedure for Type C

Suppose the data consist of pairs of measurements, and we wish to measure the strength of the association between these measurements. Correlation analysis, given in **23**, *may* be appropriate. Single large samples may be dealt with according to **24**, and single small samples by **25** and **26**. Two small samples may be compared by the method of **27**.

When there are several variables we are likely to need *partial* correlation coefficients: see **28**.

If, on the other hand, we wish to know how changes in one (dependent) variable are related to changes in other (independent) variables, regression analysis is called for. With only one independent variable, the regression coefficient is calculated as in **29**. A single large sample may be examined by **30**; a single small sample by **32**. Two samples can be compared, by **31** if large and by **33** if small.

When there is more than one independent variable, we have *multiple* regression: see **34**.

In certain circumstances, where the underlying distribution is unknown or measurements are very subjective, we may be able to assign *ranks* to the observations. A useful distribution-free measure of association, with an appropriate significance test is given in **41**.

Procedure for Type D (experimental design)

We are here concerned with specially designed experiments or with observations closely conforming to the required conditions.

To compare the means of several groups of measurements, consider the completely randomised design of **35**. The standard design requires the variances to be homogeneous: see the test in **37**.

If rather variable experimental units can be grouped into several homogeneous blocks to each of which all treatments can be applied, the use of a randomised block design is often preferable: see **36**.

Where several different factors are involved, each with two or more levels, the possibility of a factorial design should be considered. Consult Chapter 12 for general principles.

GUIDE TO FORMULAE

1. Formula for the 'normal' (or Gaussian) distribution (See section 2.1)

The normal curve is

$$y = \frac{1}{\sigma\sqrt{(2\pi)}} \exp\left\{-\frac{(x-\mu)^2}{2\sigma^2}\right\}, \qquad (1)$$

where μ is the mean and σ is the standard deviation. The variable x is continuous and may take any value from $-\infty$ to $+\infty$. The probability that x lies between any two values, say x_1 and x_2, is the area under the curve lying between these two values.

2. Calculation of mean and variance of a sample (see section 2.5)

Suppose a sample consists of n observations $x_1, x_2, \ldots, x_n$. The sample *mean*, or *average*, is given by

$$\bar{x} = \frac{1}{n}\sum x, \qquad (4)$$

where $\sum x$ means that all observations in the sample are to be added together, repeated values being included as many times as they appear.

(More elaborately, we could write $\sum x$ as

$$x_1 + x_2 + \ldots + x_n \text{ or } \sum_{i=1}^{n} x_i.)$$

The quantity $\bar{x}$ is often used to estimate the unknown population mean μ. The sample *variance* is given by

$$s^2 = \frac{1}{n-1}\sum(x-\bar{x})^2, \qquad (5)$$

where the summation on the right-hand side of formula (5) is best calculated using

$$\sum(x - \bar{x})^2 = \sum x^2 - \frac{1}{n}\left(\sum x\right)^2. \tag{7}$$

It is important to notice the divisor $n - 1$ in formula (5).

If the true population mean were known to be μ, the variance would be estimated from the sample by

$$s^2 = \frac{1}{n}\sum(x - \mu)^2. \tag{6}$$

The *standard deviation* of the sample, s, is simply the square root of the variance. We frequently use s to estimate the unknown population standard deviation σ.

If a variance is calculated from a sample in which an originally continuous variable has been grouped into intervals of constant width h, it is often best to use Sheppard's correction (see section 2.5). We adjust the estimated variance s^2, as defined above, by subtracting $\frac{1}{12}h^2$, i.e.

$$s'^2 = s^2 - \tfrac{1}{12}h^2. \tag{8}$$

3. Standard error of a mean for large samples of size $n > 30$ (see section 3.2)

Suppose the population mean is μ and the variance is σ^2. Then in a sample of size n the variance of the mean $\bar{x}$ is σ^2/n, and may be estimated by s^2/n. The standard deviation of the mean is often called its *standard error*. The sample mean is often written, together with its estimated standard error, as

$$\bar{x} \pm s/\sqrt{n}. \tag{9}$$

Care should be taken to indicate that the figure given after the $\pm$ sign is a standard error and not the estimated standard deviation s.

Standard errors are of greatest use when the distribution of the estimate to which they are attached is approximately normal. In this case a confidence interval for the unknown population mean μ is, in large samples,

$$\bar{x} - ds/\sqrt{n} \text{ to } \bar{x} + ds/\sqrt{n}, \tag{10}$$

where d is chosen from Appendix 1 to correspond to the required probability. Thus a 95 per cent range excludes 5 per cent of probability, for which d is 1.96.

4. Comparing the mean of a single large sample of size $n > 30$ with a known standard, assuming a normal population (see section 5.1)

The mean and standard deviation of a sample of size n are $\bar{x}$ and s respectively. We want to test whether the observed mean $\bar{x}$ differs significantly from a hypothetical value μ. Calculate

$$d = \frac{\bar{x} - \mu}{s/\sqrt{n}}, \tag{13}$$

where d is a normal variable with zero mean and unit standard deviation. Significance is judged by reference to Appendix 1. Thus for a two-sided test the result is significant at the 5 per cent level if the absolute value of d (i.e. without regard for sign) is greater than 1.96.

Confidence limits for the unknown population mean μ are

$$\bar{x} - ds/\sqrt{n} \text{ to } \bar{x} + ds/\sqrt{n}, \tag{14}$$

where the multiplying factor d is chosen from Appendix 1 according to the probability required. Thus if $d = 1.96$ just 5 per cent of the total frequency is excluded, and we have a 95 per cent confidence range.

It is worth noticing that if σ happens to be known exactly, we can use it instead of s in formulae (13) and (14), which are then valid even for small n.

5. Comparing the means of two large samples from normal populations (see section 5.2)

We now have two large samples. In the first there are n_1 observations, the mean is $\bar{x}_1$ and the standard deviation is s_1. The corresponding quantities in the second sample are n_2, $\bar{x}_2$ and s_2. To compare the two samples, i.e. to test whether the true means μ_1 and μ_2 are equal, calculate

$$d = \frac{\bar{x}_1 - \bar{x}_2}{\sqrt{\left(\frac{s_1^2}{n_1} + \frac{s_2^2}{n_2}\right)}} \qquad (15)$$

and refer d to Appendix 1 as before.

Confidence limits for the *difference*, $\mu_1 - \mu_2$, between the two true means are

$$(\bar{x}_1 - \bar{x}_2) - d\sqrt{\left(\frac{s_1^2}{n_1} + \frac{s_2^2}{n_2}\right)} \text{ to } (\bar{x}_1 - \bar{x}_2) + d\sqrt{\left(\frac{s_1^2}{n_1} + \frac{s_2^2}{n_2}\right)}, \qquad (16)$$

where the multiplying factor d is chosen from Appendix 1 according to the probability required.

6. Comparing the mean of a single small sample of size $n \leqslant 30$ with a known standard, assuming a normal population (see section 6.2)

The sample of n observations has mean $\bar{x}$ and standard deviation s. We want to compare $\bar{x}$ with a hypothetical mean μ, but the true standard deviation σ is unknown. Calculate the 'Student's' t with $n - 1$ degrees of freedom given by

$$t = \frac{\bar{x} - \mu}{s/\sqrt{n}}, \qquad (20)$$

and refer to Appendix 2.

Note that when this test is applied to 'paired-comparison' tests μ will usually be zero.

Confidence limits for the unknown population mean μ must also be based on 'Student's' t, and are

$$\bar{x} - ts/\sqrt{n} \text{ to } \bar{x} + ts/\sqrt{n}, \qquad (23)$$

where t, with $n - 1$ degrees of freedom, can be read from Appendix 2 according to the probability required. Thus, for 95 per cent limits with $n = 6$, i.e. 5 degrees of freedom, $t = 2.571$.

7. Comparing the means of two small samples from normal populations (see sections 6.3, 6.5)

There are two samples. In the first we have n_1 observations with mean $\bar{x}_1$ and standard deviation s_1. The corresponding quantities in the second sample are n_2, $\bar{x}_2$ and s_2. We wish to test whether the true means, μ_1 and μ_2, may be taken as equal. The method depends on whether we can assume the unknown variances, $\sigma_1{}^2$ and $\sigma_2{}^2$, to be the same or not. This point can be examined by the variance-ratio test described below. (See also section 6.5.)

Calculate the variance ratio

$$F = \frac{s_1{}^2}{s_2{}^2}, \tag{25}$$

where the samples are labelled so that $s_1{}^2$ is greater than $s_2{}^2$. We then refer to tables of the variance ratio (e.g. Appendix 5 in this book, Fisher & Yates' *Statistical Tables* or *Biometrika Tables for Statisticians*), to find the appropriate value of F for the required level of significance corresponding to $f_1 = n_1 - 1$ degrees of freedom in the numerator and $f_2 = n_2 - 1$ in the denominator.

Note that in a straightforward comparison of two samples we usually require a *two-sided* test and must enter the tables at *half* the chosen probability. In applications to the analysis of variance, on the other hand, we normally need a one-sided test, and therefore use the significance levels as they stand.

(a) *Unknown variances $\sigma_1{}^2$ and $\sigma_2{}^2$ assumed to be equal* (see section 6.3)

We calculate the 'Student's' t with $n_1 + n_2 - 2$ degrees of freedom given by

$$t = \frac{\bar{x}_1 - \bar{x}_2}{s\sqrt{\left(\dfrac{1}{n_1} + \dfrac{1}{n_2}\right)}}, \tag{21}$$

where

$$s^2 = \frac{\sum_1(x - \bar{x}_1)^2 + \sum_2(x - \bar{x}_2)^2}{n_1 + n_2 - 2}, \tag{22}$$

and refer to Appendix 2.

Confidence limits for the *difference*, $\mu_1 - \mu_2$, between the two true means are

$$(\bar{x}_1 - \bar{x}_2) - ts\sqrt{\left(\frac{1}{n_1} + \frac{1}{n_2}\right)} \text{ to } (\bar{x}_1 - \bar{x}_2) + ts\sqrt{\left(\frac{1}{n_1} + \frac{1}{n_2}\right)}, \tag{24}$$

where t, with $n_1 + n_2 - 2$ degrees of freedom, can be read from Appendix 2 according to the probability required.

(b) *Unknown variance σ_1^2 and σ_2^2 not assumed to be equal* (see section 6.5)
Use formula (15) above, i.e.

$$d = \frac{\bar{x}_1 - \bar{x}_2}{\sqrt{\left(\frac{s_1^2}{n_1} + \frac{s_2^2}{n_2}\right)}}, \tag{15}$$

but now treat this as being distributed approximately like 'Student's' t with f degrees of freedom, the latter being given by

$$\left.\begin{aligned} f &= \frac{1}{\frac{u^2}{n_1 - 1} + \frac{(1-u)^2}{n_2 - 1}}, \\ \text{where} \quad & \\ u &= \frac{s_1^2/n_1}{s_1^2/n_1 + s_2^2/n_2}. \end{aligned}\right\} \tag{26}$$

Confidence limits for the difference, $\mu_1 - \mu_2$, between the two true means are given by formula (16) in **5** above, remembering that we are here treating d as a t with f degrees of freedom. Note that non-integral values of f must be treated by appropriate interpolation.

8. Formula for the binomial distribution (see section 2.2)

The chance that an individual has a certain character is $p(=1-q)$. The probability that a sample of n independent individuals contains exactly a with the character is

$$\frac{n!}{a!(n-a)!}p^a q^{n-a}, \qquad (2)$$

where a may be any whole number from 0 to n. The factorial $x!$ means $1.2.3 \ldots (x-1)x$, and $0!$ and $1!$ are both taken to be unity. The mean of the distribution of a is np, and the variance is npq (see section 2.5). The probability p may be estimated from the sample by a/n, which has mean p and variance pq/n.

9. Normal approximation to binomial distribution in large samples of size $n > 30$ (see section 3.3)

When samples are large, the binomial distribution tends to normality. We can then regard the estimate a/n of the unknown probability p as approximately normal, and write it with its standard error as

$$\frac{a}{n} \pm \sqrt{\left[\frac{\frac{a}{n}\left(1-\frac{a}{n}\right)}{n}\right]}. \qquad (11)$$

Confidence limits can then be calculated as for normal distributions. (If samples are small, or if a/n is near 0 or 1, confidence limits can be obtained from *Biometrika Tables for Statisticians*.)

10. Comparison of a percentage based on a large sample of size $n > 30$ with a known standard (see section 5.3)

Suppose that a sample of n individuals contains a with a certain character. We want to know whether the estimated chance a/n of an individual having the character is significantly different from a hypothetical value p. Calculate

$$d = \frac{(a/n) - p}{\sqrt{\left(\frac{p(1-p)}{n}\right)}}, \qquad (17)$$

and refer d to Appendix 1 as before.

11. Comparison of two percentages based on two large samples (see section 5.3)

We have two samples, one of n_1 individuals with a_1 having a certain character, the other of n_2 individuals with a_2 having the character. We want to know whether the two samples may be considered as drawn from the same population, i.e. whether the true probability p of having the character is the same for both samples.

First, calculate

$$k_1 = \frac{a_1}{n_1};\ k_2 = \frac{a_2}{n_2};\ k = \frac{a_1 + a_2}{n_1 + n_2}.$$

Then find

$$d = \frac{k_1 - k_2}{\sqrt{\left[k(1-k)\left(\frac{1}{n_1} + \frac{1}{n_2}\right)\right]}}, \tag{18}$$

and refer d to Appendix 1 as before.

Confidence limits for the difference $p_1 - p_2$ between the two unknown percentages are

$$(k_1 - k_2) - d\sqrt{\left[\frac{k_1(1-k_1)}{n_1} + \frac{k_2(1-k_2)}{n_2}\right]}$$

to

$$(k_1 - k_2) + d\sqrt{\left[\frac{k_1(1-k_1)}{n_1} + \frac{k_2(1-k_2)}{n_2}\right]}, \tag{18a}$$

where d is chosen from Appendix 1 to correspond to the required probability.

12. Formula for the Poisson distribution (see section 2.3)

If the total number of individuals observed under certain circumstances varies according to a Poisson distribution with mean m, the chance of obtaining exactly x is

$$\frac{e^{-m} m^x}{x!}, \tag{3}$$

where x can be any whole number from 0 to ∞. (The factorial $x!$ is defined above in **8**.) The variance of the distribution of x is equal to the mean m (see section 2.5). We can use the mean $\bar{x}$ of several observed values as an estimate of the unknown m.

13. Normal approximation to Poisson distribution when the product nx is large, e.g. > 30 (see section 3.4)

Suppose we have n observations each having a Poisson distribution with true mean m. If the product $n\bar{x}$ is large, e.g. greater than 30, the distribution of $\bar{x}$ will be approximately normal. The estimate of m, with its standard error can then be written as

$$\bar{x} \pm \sqrt{\frac{\bar{x}}{n}}. \tag{12}$$

Confidence limits can be calculated as for normal distributions. (For small values of Poisson variables, confidence limits can be obtained from *Biometrika Tables for Statisticians*.)

14. Comparing two Poisson distributions (see section 5.3)

Suppose we have two samples of observations. The first sample has n_1 observations, each following the same Poisson distribution, with mean $\bar{x}_1$. The second sample of n_2 observations from a Poisson population has mean $\bar{x}_2$. We test whether the two Poisson samples are significantly different by calculating

$$d = \frac{\bar{x}_1 - \bar{x}_2}{\sqrt{\left(\frac{\bar{x}_1}{n_1} + \frac{\bar{x}_2}{n_2}\right)}}, \tag{19}$$

and referring d to Appendix 1 as before. For this test to be reasonably reliable, it is as well to have $n_1\bar{x}_1$ and $n_2\bar{x}_2$ each larger than 30.

Confidence limits for the difference $m_1 - m_2$ between the two unknown Poisson parameters are

$$(\bar{x}_1 - \bar{x}_2) - d\sqrt{\left(\frac{\bar{x}_1}{n_1} + \frac{\bar{x}_2}{n_2}\right)} \quad \text{to} \quad (\bar{x}_1 - \bar{x}_2) + d\sqrt{\left(\frac{\bar{x}_1}{n_1} + \frac{\bar{x}_2}{n_2}\right)}, \tag{19a}$$

where d is chosen from Appendix 1 to correspond to the required probability.

(To compare two small single observations we can refer to *Biometrika Tables for Statisticians*. Note that the table given is 'one-tailed' as it stands.)

15. Goodness-of-fit χ^2 (see sections 7.1, 8.1)

When data can be broken down into a set of compartments, for each of which there is an observed number (O) and an expected number (E) of individuals, the goodness-of-fit χ^2 is given by

$$\chi^2 = \sum \frac{(O - E)^2}{E}, \tag{27}$$

where the summation is over all compartments. Care must be taken to use the right number of degrees of freedom. For a contingency table of r rows and c columns this is $(r - 1)(c - 1)$. When fitting a frequency distribution to data grouped into k classes, the number of degrees of freedom is $k - 1$ minus the number of constants estimated from the data (e.g. two for the normal distribution).

In general, every additional independent constraint removes one degree of freedom.

16. General contingency table with r rows and c columns (see section 7.1)

The table with marginal and grand totals may be written as in Table 6. For the top left-hand entry we have $O = a$ and $E = AB/N$. Similarly, to obtain the expected value for any other entry, we multiply the corresponding marginal totals and divide by the grand total. The appropriate value of χ^2 is then calculated from formula (27) in **15** above and has $(r - 1)(c - 1)$ degrees of

Table 6

a c $\vdots$	b d $\vdots$	$\dots$ $\dots$ $\vdots$	A $\cdot$
B	$\cdot$	$\dots$	N

freedom. The method is valid only if no expected number is less than about 5.

17. Contingency table with only 2 rows and c columns (see section 7.2)

The table is shown in Table 8.

Table 8

$\dots$ $\dots$	a b	$\dots$ $\dots$	A B
$\dots$	n	$\dots$	N

Brandt & Snedecor's formula for χ^2 is

$$\chi^2 = \frac{\sum \frac{a^2}{n} - \frac{A^2}{N}}{k(1-k)} = \frac{\sum \frac{b^2}{n} - \frac{B^2}{N}}{k(1-k)}, \tag{28}$$

where $k = A/N$, $1 - k = B/N$, and there are $c - 1$ degrees of freedom. The method is valid only if no expected number is less than about 5.

18. 2×2 contingency tables (see section 7.3)

The table is shown in Table 9.

Table 9

a c	b d	$a + b$ $c + d$
$a + c$	$b + d$	n

The formula for χ^2 with Yates' correction is

$$\chi^2 = \frac{n\{|ad - bc| - \frac{1}{2}n\}^2}{(a + b)(c + d)(a + c)(b + d)}, \tag{30}$$

where the vertical lines in $|ad - bc|$ mean that we must take the absolute, i.e. positive, value of the difference between ad and bc. There is just one degree of freedom, so that, e.g. the 5 per cent point is 3.84 for a two-tailed test. The method is valid only if no expected number is less than about 5.

19. Exact test for 2 × 2 contingency table (see section 7.4)

If any expectation in a 2 × 2 table is less than about 5, we cannot use the χ^2 method. We must then calculate the probability of the observed table and of all other equally or more extreme tables, if any, remembering to include tables from both extremes if a two-tailed test is appropriate. The sum of these probabilities gives the significance level P achieved.

The probability of Table 9 in **18** is

$$\frac{(a + b)!(c + d)!(a + c)!(b + d)!}{n!a!b!c!d!}, \tag{31}$$

where the factorial $x!$ means $1.2.3 \ldots (x - 1)x$, and 0! and 1! are both taken to be unity. If the expression in formula (31) contains any numbers that are at all large, it is best evaluated using logarithms. Logarithms of all factorials up to 150! are given in Fisher & Yates' *Statistical Tables*, and up to 1000! in *Biometrika Tables for Statisticians*.

20. Homogeneity test for several binomial samples (see section 8.4)

Suppose we have several independent batches of data, each classified into two groups whose frequencies follow some binomial distribution. Homogeneity can be tested by examining the resultant contingency table with only *two* rows (see **17**). If only two samples are to be compared, we have a 2 × 2 table (see **18**); but when numbers are very small, we must resort to the 'exact' test (see **19**).

21. Homogeneity test for several Poisson samples each with a single observation (see section 8.4)

Suppose we have k samples, each of which yields a single number of individuals represented by x, where x is assumed to have a Poisson distribution. A suitable test of homogeneity is given by

$$\chi^2 = \frac{\sum(x - \bar{x})^2}{\bar{x}} = \frac{1}{\bar{x}}\left\{\sum x^2 - \frac{1}{k}\left(\sum x\right)^2\right\}, \qquad (32)$$

where the number of degrees of freedom is $k - 1$. To be on the safe side $\bar{x}$ should be not less than about 5 units.

22. Additive property of χ^2 and normal approximation (see section 8.5)

The sum of several χ^2's is also a χ^2 whose degrees of freedom are given by adding the individual degrees of freedom for each component. This may be useful in combining results from several tests, but often leads to a χ^2 with a large number of degrees of freedom, say f, not covered by Appendix 3. We can, however, use the result that

$$d = \sqrt{(2\chi^2)} - \sqrt{(2f - 1)} \qquad (33)$$

is approximately normal with zero mean and unit standard deviation. Significance is then tested by reference to Appendix 1, remembering to follow the rule for a one-tailed test.

23. Calculation of the correlation coefficient (see section 9.2)

The basic observations are n pairs of associated observations represented by (x, y), and we assume that x and y follow, at least approximately, a bivariate normal distribution. The calculations required are best effected by proceeding systematically as follows.

First, find the sums, and sums of squares and products given by the scheme

$$\left.\begin{array}{lll} n, & & \\ \sum x, & \sum y, & \\ \sum x^2, & \sum y^2, & \sum xy. \end{array}\right\} \tag{35}$$

Remember that the values of some items may be repeated, especially if calculations are made on grouped data, and must be counted as many times as they occur. Next calculate the three quantities in the first line of formula (37) below, and subtract each of these from the corresponding quantities in the last line of formula (35) to give the second line of formula (37):

$$\left.\begin{array}{lll} \frac{1}{n}\left(\sum x\right)^2, & \frac{1}{n}\left(\sum y\right)^2, & \frac{1}{n}\left(\sum x\right)\left(\sum y\right), \\ \sum(x-\bar{x})^2, & \sum(y-\bar{y})^2, & \sum(x-\bar{x})(y-\bar{y}). \end{array}\right\} \tag{37}$$

Divide through the second line of formula (37) by $n - 1$ to give the two estimated variances and the estimated *covariance*:

$$s_x^2, s_y^2, c. \tag{38}$$

We now have all the basic quantities required. The estimate of the unknown population correlation coefficient ρ is

$$\left.\begin{aligned} r &= \frac{c}{s_x s_y} \\ &= \frac{\sum(x-\bar{x})(y-\bar{y})}{\sqrt{\left[\sum(x-\bar{x})^2\sum(y-\bar{y})^2\right]}}. \end{aligned}\right\} \tag{39}$$

We can, if we wish, adjust the two variances (but not the covariance), by Sheppard's corrections and use the first line of formula (39), with adjusted standard deviations, to give the best estimate of ρ. For significance tests Sheppard's corrections are omitted, and we can use the second line of formula (39) directly.

24. Normal approximation to the distribution of the correlation coefficient in very large samples of size $n \geqslant 500$ (see section 9.3)

If n is greater than about 500, r is approximately normally distributed with mean ρ and standard deviation $(1 - \rho^2)/\sqrt{n}$. We can then use the estimate with attached standard error given by

$$r \pm \frac{1 - r^2}{\sqrt{n}}. \tag{40}$$

Significance is then tested or confidence limits are set as usual for normal variables.

25. Exact t-test for the existence of correlation in small samples (see section 9.3)

To test whether there is any correlation at all, i.e. whether $\rho = 0$, even for small sample sizes, we calculate

$$t = \frac{r\sqrt{(n - 2)}}{\sqrt{(1 - r^2)}}. \tag{41}$$

This is then a 'Student's' t-distribution with $n - 2$ degrees of freedom, and significance is tested accordingly. Percentage points may, however, be read directly from Appendix 4.

26. Comparing the correlation coefficient, based on a single small sample, with a non-zero standard (see section 9.3)

When testing the significance of departures of r from non-zero values of ρ, we use the transformation

$$\left.\begin{aligned} z &= \tfrac{1}{2}\log_e \frac{1 + r}{1 - r}, \\ \zeta &= \tfrac{1}{2}\log_e \frac{1 + \rho}{1 - \rho}. \end{aligned}\right\} \tag{42}$$

The difference $z - \zeta$ is then normally distributed with zero mean and variance $1/(n - 3)$. Significance is tested as for normal variables.

Notice that the logarithms on the right of formula (42) are *natural* logarithms. We can, however, obtain z (or ζ) directly, given r (or ρ), by reference to *Biometrika Tables for Statisticians*.

Confidence limits for ρ are obtained by first finding confidence limits for ζ and then transforming back.

27. Comparing the correlation coefficients calculated from two small samples (see section 9.3)

Suppose we have two small samples, one with n_1 observations, the other with n_2. We want to compare the two estimated correlation coefficients r_1 and r_2. This is done by using formula (42) in **26** to calculate z_1 and z_2 for the two samples. We then use the result that $z_1 - z_2$ is normally distributed with zero mean and variance $1/(n_1 - 3) + 1/(n_2 - 3)$. Significance is tested as for normal variables.

Confidence limits for the unknown difference $\rho_1 - \rho_2$ are obtained by first finding the limits for $\zeta_1 - \zeta_2$ and then transforming back.

28. Calculation of partial correlation coefficients (see section 14.2)

If there are more than two basic variables, we may wish to examine the correlation between any pair when the effect of one or more of the remaining variables has been eliminated. Thus, with three variables, represented by the numbers 1, 2 and 3, we estimate the *partial correlation* between 1 and 2, eliminating 3, by $r_{12.3}$. This can be calculated in terms of the total correlation coefficients r_{12}, r_{13} and r_{23}, by means of

$$r_{12.3} = \frac{r_{12} - r_{13}r_{23}}{\sqrt{[(1 - r_{13}^2)(1 - r_{23}^2)]}}. \qquad (67)$$

There are similar formulae for $r_{13.2}$ and $r_{23.1}$. We also have analogous expressions to formula (67) for the true correlation coefficients like $\rho_{12.3}$, etc.

With four variables represented by 1, 2, 3 and 4, we might want to go a stage further and eliminate the effects of both 3 and 4. This is done by calculating the coefficient $r_{12.34}$ given by

$$r_{12.34} = \frac{r_{12.4} - r_{13.4}r_{23.4}}{\sqrt{[(1 - r^2_{13.4})(1 - r^2_{23.4})]}}, \tag{68}$$

where coefficients on the right-hand side like $r_{12.4}$ must be calculated from formulae like formula (67).

Partial correlation coefficients can be tested for significance in much the same way as total correlation coefficients. We simply use formulae like formulae (40), (41) and (42) in **24**, **25** and **26** above, with the modification that n is replaced by n minus as many variables have been eliminated from the comparison in question. For such significance tests to be valid, variables *not* eliminated from any comparison should follow a multivariate normal distribution.

29. Calculation of regression coefficients (see section 10.2)

The basic observations are n pairs of associated observations represented by (x, y). Although we are here concerned with variations in one variable, say y, relative to variations in the other, say x, the initial calculations are exactly the same as for correlation: see formulae (35), (37) and (38) in **23** above.

The true regression line for the regression of y on x is

$$y = \alpha + \beta x. \tag{43}$$

We assume that the distribution of y about the regression line is normal with zero mean and variance σ^2 independent of x. No assumption is made about the distribution of x. We estimate the true regression coefficient β by

$$\left.\begin{aligned} b &= \frac{\sum(x - \bar{x})(y - \bar{y})}{\sum(x - \bar{x})^2} \\ &= \frac{c}{s_x^2} \end{aligned}\right\} \tag{44}$$

and the constant α by

$$a = \bar{y} - b\bar{x}. \tag{45}$$

The *fitted* regression line is now

$$y = a + bx. \tag{46}$$

The variance σ^2 of the deviations of y from the regression line is estimated by

$$s^2 = \frac{1}{n-2}\left\{\sum(y-\bar{y})^2 - \frac{\left[\sum(x-\bar{x})(y-\bar{y})\right]^2}{\sum(x-\bar{x})^2}\right\}. \tag{47}$$

30. Normal approximation to the distribution of regression coefficients based on large samples of size $n > 30$ (see section 10.3)

The estimated regression coefficient b is distributed normally with mean β and variance $\sigma^2/\sum(x-\bar{x})^2$. In large samples we can use the estimate with attached standard error given by

$$b \pm \frac{s}{\sqrt{\left[\sum(x-\bar{x})^2\right]}}. \tag{48}$$

Significance is then tested or confidence limits are set as usual for a normal variable.

31. Comparing the regression coefficients calculated from two large samples (see section 10.3)

If both samples are large we calculate

$$d = \frac{b_1 - b_2}{\sqrt{\left[\frac{{s_1}^2}{\sum_1(x-\bar{x}_1)^2} + \frac{{s_2}^2}{\sum_2(x-\bar{x}_2)^2}\right]}}, \tag{50}$$

and refer d to Appendix 1. Significance is then tested or confidence limits are set as usual for normal variables.

32. Comparing the regression coefficient, calculated from a single small sample of size $n \leqslant 30$, with a known standard (see section 10.3)

In small samples we calculate the 'Student's' t as

$$t = \frac{b - \beta}{s\Big/\sqrt{\left[\sum(x - \bar{x})^2\right]}}, \tag{49}$$

where there are $n - 2$ degrees of freedom.

Departures from a hypothetical value of the regression coefficient β, which will be zero if we are examining the existence of regression, are then tested as usual by means of the t-distribution.

Confidence limits for β are also obtained in the same way as for t-variables.

33. Comparing the regression coefficients calculated from two small samples (see section 10.3)

The test to be used depends on whether the two residual variances, σ_1^2 and σ_2^2, can be assumed equal or not. We examine this point by means of the variance-ratio $F = s_1^2/s_2^2$. See formula (25) *et seq*. in **7** above.

(a) *Unknown residual variances σ_1^2 and σ_2^2 assumed to be equal* (see section 10.3)
We calculate the 'Student's' t

$$\left.\begin{aligned} t &= \frac{b_1 - b_2}{s\sqrt{\left[\dfrac{1}{\sum_1(x - \bar{x}_1)^2} + \dfrac{1}{\sum_2(x - \bar{x}_2)^2}\right]}}, \\ &\text{where} \\ s^2 &= \frac{(n_1 - 2)s_1^2 + (n_2 - 2)s_2^2}{n_1 + n_2 - 4}, \end{aligned}\right\} \tag{51}$$

and the number of degrees of freedom is $n_1 + n_2 - 4$. Significance is tested and confidence limits are set as usual for the t-distribution.

(b) *Unknown residual variances σ_1^2 and σ_2^2 not assumed to be equal* (see section 10.3)
We calculate d as in formula (50) above, i.e.

$$d = \frac{b_1 - b_2}{\sqrt{\left[\dfrac{s_1^2}{\sum_1(x-\bar{x}_1)^2} + \dfrac{s_2^2}{\sum_2(x-\bar{x}_2)^2}\right]}}, \tag{50}$$

but treat it as a 'Student's' t with f degrees of freedom given by

$$f = \frac{1}{\dfrac{u^2}{n_1 - 2} + \dfrac{(1-u)^2}{n_2 - 2}},$$

where

$$u = \frac{s_1^2/\sum_1(x-\bar{x}_1)^2}{s_1^2/\sum_1(x-\bar{x}_1)^2 + s_2^2/\sum_2(x-\bar{x}_2)^2}. \tag{52}$$

Confidence limits for the difference $\beta_1 - \beta_2$ are obtained as usual with t-variables.

34. Calculation of partial regression coefficients (see section 14.3)

(a) *Two independent variables only* (see section 14.3)
We wish to examine the relation of a single dependent variable y to *two* independent variables x_1 and x_2. Suppose that the mean value of y, given x_1 and x_2, is

$$y = \alpha + \beta_1 x_1 + \beta_2 x_2. \tag{69}$$

We assume that the distribution of y, given x_1 and x_2, is normal with the mean given by formula (69) and variance σ^2 independent of x_1 and x_2.

We estimate the *partial regression coefficients*, β_1 and β_2, by means of b_1 and b_2, whose calculation is described below. The fitted regression is then

$$y = a + b_1x_1 + b_2x_2, \tag{70}$$

where

$$a = \bar{y} - b_1\bar{x}_1 - b_2\bar{x}_2. \tag{71}$$

We start by calculating the sums, and sums of squares and products, according to the scheme

$$\left.\begin{array}{llll} n, & \sum x_1, & \sum x_2, & \sum y, \\ & \sum x_1^2, & \sum x_1x_2, & \sum x_1y, \\ & & \sum x_2^2, & \sum x_2y, \\ & & & \sum y^2. \end{array}\right\} \tag{72}$$

Next, we require the 'correction-factors'

$$\left.\begin{array}{lll} \frac{1}{n}\left(\sum x_1\right)^2, & \frac{1}{n}\left(\sum x_1\right)\left(\sum x_2\right), & \frac{1}{n}\left(\sum x_1\right)\left(\sum y\right), \\ & \frac{1}{n}\left(\sum x_2\right)^2, & \frac{1}{n}\left(\sum x_2\right)\left(\sum y\right), \\ & & \frac{1}{n}\left(\sum y\right)^2, \end{array}\right\} \tag{74}$$

which are to be subtracted from the corresponding quantities in the last three lines of formula (72) to give

$$\left.\begin{array}{lll} A, & B, & D, \\ & C, & E, \\ & & S, \end{array}\right\} \tag{75}$$

i.e. $A = \sum x_1^2 - (1/n)(\sum x_1)^2$, etc. Then

$$\left.\begin{array}{l} b_1 = \dfrac{CD - BE}{AC - B^2}, \\ b_2 = \dfrac{AE - BD}{AC - B^2}. \end{array}\right\} \tag{76}$$

The residual variance σ^2 is estimated by

$$s^2 = \frac{S - b_1 D - b_2 E}{n - 3}. \tag{77}$$

The standard errors of the two estimated regression coefficients are

$$s_{b_1} = \frac{s}{\sqrt{\left(\frac{AC - B^2}{C}\right)}},$$

and (78)

$$s_{b_2} = \frac{s}{\sqrt{\left(\frac{AC - B^2}{A}\right)}}.$$

Significance tests are performed and confidence limits are set as for ordinary regression coefficients, but remember that the t-variables here have $n - 3$ degrees of freedom.

(b) *More than two independent variables* (see section 14.4)
We illustrate the general method by the case of three independent variables, x_1, x_2 and x_3. Suppose that the mean value of y, given x_1, x_2 and x_3, is

$$y = \alpha + \beta_1 x_1 + \beta_2 x_2 + \beta_3 x_3, \tag{79}$$

while the fitted regression is

$$y = a + b_1 x_1 + b_2 x_2 + b_3 x_3, \tag{80}$$

where

$$a = \bar{y} - b_1 \bar{x}_1 - b_2 \bar{x}_2 - b_3 \bar{x}_3. \tag{81}$$

We assume that the distribution of y, given x_1, x_2 and x_3, is normal with the mean given by formula (79) and variance σ^2 independent of x_1, x_2 and x_3.

Sums of squares and products about mean values are calculated as for the case of only two independent variables. Let us write these sums of squares and products about mean values as

$$\left.\begin{aligned}&\sum(y-\bar{y})^2 = c_{00},\\&\sum(x_i-\bar{x}_i)(y-\bar{y}) = c_{0i};\ i=1,2,3,\\&\sum(x_i-\bar{x}_i)(x_j-\bar{x}_j) = c_{ij};\ i,j=1,2,3.\end{aligned}\right\} \quad (82)$$

Now consider the *three sets* of equations in X_1, X_2 and X_3, given by

$$\begin{aligned}c_{11}X_1 + c_{12}X_2 + c_{13}X_3 &= \left.\begin{matrix}1\\0\\0\end{matrix}\right\},\left.\begin{matrix}0\\1\\0\end{matrix}\right\},\left.\begin{matrix}0\\0\\1\end{matrix}\right\}.\\ c_{12}X_1 + c_{22}X_2 + c_{23}X_3 &\\ c_{13}X_1 + c_{23}X_2 + c_{33}X_3 &\end{aligned} \quad (83)$$

Find the three sets of solutions

$$\begin{matrix}X_1 = \\ X_2 = \\ X_3 = \end{matrix}\left.\begin{matrix}a_{11}\\a_{12}\\a_{13}\end{matrix}\right\},\left.\begin{matrix}a_{12}\\a_{22}\\a_{23}\end{matrix}\right\},\left.\begin{matrix}a_{13}\\a_{23}\\a_{33}\end{matrix}\right\}. \quad (84)$$

The required partial regression coefficients are now

$$\left.\begin{aligned}b_1 &= a_{11}c_{01} + a_{12}c_{02} + a_{13}c_{03},\\ b_2 &= a_{12}c_{01} + a_{22}c_{02} + a_{23}c_{03},\\ b_3 &= a_{13}c_{01} + a_{23}c_{02} + a_{33}c_{03}.\end{aligned}\right\} \quad (85)$$

The estimate of the residual variance, σ^2, is

$$s^2 = \frac{c_{00} - b_1c_{01} - b_2c_{02} - b_3c_{03}}{n-4}. \quad (86)$$

The standard errors of the three partial regression coefficients are

$$s_{b_1} = s\sqrt{a_{11}},\quad s_{b_2} = s\sqrt{a_{22}},\quad s_{b_3} = s\sqrt{a_{33}}. \quad (87)$$

In applications to larger numbers of independent variables, say k in all, there will be further sums of squares and products in formula (82); formula (83) will have k sets of k equations in k unknowns; formula (84) will be extended to k columns and k rows, and formula (85) to k rows; while formula (86) will have k negative terms in the numerator and a denominator of $n - k - 1$, the latter being the appropriate number of degrees of freedom for t-tests.

See also the end of section 14.4 for a more streamlined version, using elementary matrix algebra suitable for the automatic handling of formulae (82)–(87) by a good pocket calculator or a computer with appropriate software.

35. Completely randomised designs (see section 11.2)

The data appear as in Table 15. Calculate

$$\left.\begin{aligned}\sum(x_{ij}-\bar{x})^2 &= \sum x_{ij}^2 - C,\\ \text{where}\qquad C &= G^2/N,\end{aligned}\right\} \tag{53}$$

and

$$\sum T_i^2/n_i - C. \tag{54}$$

The summation in formula (53) is over all observations and in formula (54) over all k varieties.

The analysis of variance is shown in Table 16.

Table 15

Variety	Observed plot yields	No. of obs.	Total	Mean
1	$x_{11}, x_{12}, \ldots, x_{1n_1}$	n_1	T_1	$\bar{x}_1$
2	$x_{21}, x_{22}, \ldots, x_{2n_2}$	n_2	T_2	$\bar{x}_2$
.	.	.	.	.
.	.	.	.	.
.	.	.	.	.
i	$x_{i1}, x_{i2}, \ldots, x_{in_i}$	n_i	T_i	$\bar{x}_i$
.	.	.	.	.
.	.	.	.	.
.	.	.	.	.
k	$x_{k1}, x_{k2}, \ldots, x_{kn_k}$	n_k	T_k	$\bar{x}_k$
Total		$N = \sum n_i$	$G = \sum T_i$	$\bar{x} = G/N$

Table 16

Source of variation	Sum of squares	Degrees of freedom	Mean squares
Between varieties	$\sum T_i^2/n_i - C$	$k-1$	M
Residual	By subtraction	$N-k$	s^2
Total	$\sum x_{ij}^2 - C$	$N-1$	–

For a significance test of the difference between varieties we examine the variance-ratio

$$F = M/s^2, \tag{56}$$

with $f_1 = k - 1$ and $f_2 = N - k$ degrees of freedom in the numerator and denominator respectively. Appendix 5 of this book gives 5 per cent and 1 per cent points of the variance-ratio F. We can also refer to Fisher & Yates' *Statistical Tables* (where F is called e^{2z}), or *Biometrika Tables for Statisticians*.

We can write the mean of the ith variety together with its standard error as

$$\bar{x}_i \pm s/\sqrt{n_i}, \tag{57}$$

and the difference between the means of ith and jth varieties as

$$(\bar{x}_i - \bar{x}_j) \pm s\sqrt{\left(\frac{1}{n_i} + \frac{1}{n_j}\right)}. \tag{58}$$

Significance tests are performed and confidence limits are set by reference to the normal distribution if $N - k$ is large, and the t-distribution (with $N - k$ degrees of freedom) if $N - k$ is small.

36. Randomised block designs (see section 11.3)

The data appear as in Table 20. Calculate

$$\left.\begin{aligned} &\sum_i T_i^2/b - C, \\ \text{where} \quad & \\ &C = G^2/bt, \end{aligned}\right\} \tag{59}$$

and

$$\sum_j B_j^2/t - C. \tag{60}$$

The analysis of variance table is as in Table 21.

The block and treatment effects are tested by examining the variance-ratios M_B/s^2 and M_T/s^2.

Table 20

Treatment	Block 1	2	...	j	...	b	Treatment total	Treatment mean
1	x_{11}	x_{12}	...	x_{1j}	...	x_{1b}	T_1	$\bar{x}_1$
2	x_{21}	x_{22}	...	x_{2j}	...	x_{2b}	T_2	$\bar{x}_2$
.	.	.		.		.	.	.
.	.	.		.		.	.	.
.	.	.		.		.	.	.
i	x_{i1}	x_{i2}	...	x_{ij}	...	x_{ib}	T_i	$\bar{x}_i$
.	.	.		.		.	.	.
.	.	.		.		.	.	.
.	.	.		.		.	.	.
t	x_{t1}	x_{t2}	...	x_{tj}	...	x_{tb}	T_t	$\bar{x}_t$
Block total	B_1	B_2	...	B_j	...	B_b	G	$\bar{x} = G/bt$

Table 21

Source of variation	Sum of squares	Degrees of freedom	Mean square
Treatment	$\sum_i T_i^2/b - C$	$t - 1$	M_T
Blocks	$\sum_j B_j^2/t - C$	$b - 1$	M_B
Residual	By subtraction	$(t - 1)(b - 1)$	s^2
Total	$\sum x_{ij}^2 - C$	$bt - 1$	—

The mean of the ith treatment with its standard error is

$$\bar{x}_i \pm s/\sqrt{b}, \tag{61}$$

and the difference between any two treatments, say the ith and jth, is

$$(\bar{x}_i - \bar{x}_j) \pm s\sqrt{\frac{2}{n}}. \tag{62}$$

Significance tests are performed and confidence limits are set by reference to the normal or t-distributions as appropriate.

Missing plot technique
If a single value is accidentally lost, say that for the observation x_{ij}, we calculate an estimated value

$$x'_{ij} = \frac{tT'_i + bB'_j - G'}{(t-1)(b-1)}, \tag{63}$$

where T'_i, B'_j and G' are the treatment, block and grand totals for the observations actually available.

The analysis is then performed as though x'_{ij} were a real observation, except that both the total sum of squares and the residual sum of squares each lose 1 degree of freedom.

37. Bartlett's test for the homogeneity of several variances (see section 11.4)

Suppose we have k samples, the ith of which has n_i observations and yields an estimated variance of s_i^2. We want to test whether all these estimates are homogeneous, i.e. whether all k samples might be regarded as having the same true variance, say σ^2 (although their true means might differ).

First, calculate

$$s^2 = \frac{\sum f_i s_i^2}{f},$$

where

$$f_i = n_i - 1 \text{ and } f = \sum f_i. \tag{65}$$

Then find

$$\frac{1}{C}\left\{f \log_{10} s^2 - \sum f_i \log_{10} s_i^2\right\},$$

where

$$C = 0.4343\left[1 + \frac{1}{3(k-1)}\left\{\sum \frac{1}{f_i} - \frac{1}{f}\right\}\right]. \tag{66}$$

The quantity in the first line of formula (66) is then distributed approximately like χ^2 with $k - 1$ degrees of freedom.

The main snag about this test is that it depends on the distributions in the several samples all being normal, whether or not the variances are equal. Moderate departures from normality may cause the results of the test to be misleading.

38. Sign tests (see section 15.3)

Consider the 'paired-comparison' test discussed first in section 6.2 and again in section 15.2, and also referred to in section **6** of this summary. If there are serious doubts as to the validity of the t-test, when applied to the series of differences between members of successive pairs of observations, we can use a simple *distribution-free* test as follows. Only the *signs* of the differences are examined. The number of positive (or negative) values will follow a binomial distribution with $p = \frac{1}{2}$, and we test to see if a significantly high or low proportion has occurred, using tests referred to in sections **9** and **10** of this summary, and described in more detail in sections 3.3, 5.3 and 15.3. However, Wilcoxon's test given in section **39** is usually to be preferred.

39. Wilcoxon's signed rank sum test for a single sample (see section 15.4)

A distribution-free test that is more powerful, in the paired-comparison situation, than the sign test of section **38** is Wilcoxon's test for a single sample. We first put all the observations in ascending order of magnitude, *ignoring the signs*. Zero values are rejected altogether, and the remaining non-zero values are assigned the ranks 1 to n. If any observations are *numerically* equal (or *tied*) they are each assigned an average rank calculated from the ranks that would otherwise have been used.

Then calculate the sum T of the ranks attached to observations that are in fact positive. Appendix 6 then gives the extreme values of T, for all $n \leqslant 25$, that would have to be *attained or exceeded* for significance to occur.

For larger values of n use the approximation given by

$$d = \{|T - \tfrac{1}{4}n(n+1)| - \tfrac{1}{2}\}/\sqrt{v}, \tag{88}$$

where

$$v = n(n + 1)(2n + 1)/24, \tag{89}$$

and we can take d to be normally distributed with zero mean and unit standard deviation. When tied ranks occur, each group of t tied ranks reduces v by $(t^3 - t)/48$.

40. Wilcoxon's rank sum test for two samples (see section 15.5)

Suppose that there are two samples to be compared, comprising the n values $x_1, x_2, \ldots, x_n$ and the m values $y_1, y_2, \ldots, y_m$. We wish to make a distribution-free test of the hypothesis that the x's and y's have identical distributions, but are primarily interested in differences of location, i.e. differences between mean values.

We first combine the observations from both samples, put them in ascending order of magnitude and assign ranks, making sure that the x and y components can be easily identified later. Tied values are treated as in **39** above.

We then calculate the sum T of the ranks of all the x's. Significance is determined by reference to tables showing the extreme values of T that have to be *attained or exceeded*. (See Ciba–Geigy's *Scientific Tables*, which cover the values $n \leqslant 25$, $m \leqslant 50$. Note that in these tables $n = N_1$, $m = N_2$.)

A convenient approximation for values of n and m beyond the tables is given by

$$d = \{T - \tfrac{1}{2}n(n + m + 1)\}/\sqrt{v}, \tag{90}$$

where

$$v = nm(n + m + 1)/12, \tag{91}$$

and we can take d to be normally distributed with zero mean and unit standard deviation.

When ties occur we must write v as

$$v = \frac{nm(N^3 - N - R)}{12N(N - 1)}. \tag{92}$$

where

$$N = n + m, \tag{93}$$

and the reduction term R is found by adding together all the quantities $t^3 - t$ arising from each group of t tied ranks.

41. Kendall's rank correlation coefficient τ (see section 15.6)

Suppose that we have n individuals, each of which can be classified under each of two headings A and B. For either classification it is envisaged that individual measurements may have no objective meaning, but that the whole *series* of n measurements can at least be put into a reliable order and assigned ranks from 1 to n.

We therefore have two rankings of n individuals, and can carry out a distribution-free test of association by calculating Kendall's rank correlation coefficient τ. This is obtained as follows.

First, arrange the individual pairs in two columns so that one series, for the A-classification, say, appears in standard ascending order of magnitude from 1 to n. We then take each item of the B-classification in turn, starting at the top, and count how many rankings *below* the item in question have a higher value. The sum of all these higher rankings for all items in the column is P. A further quantity Q is obtained in a similar way by counting rankings with lower values than each item considered.

We then define S as

$$S = P - Q,$$

and Kendall's τ is given by

$$\tau = S/\{\tfrac{1}{2}n(n - 1)\} \tag{94}$$

Significance tests for small values of n, e.g. $n \leq 10$, can be performed directly on S itself by reference to Appendix 7.

When n is beyond the scope of this simple table we use an approximation given by

$$d = S/\sigma_s, \tag{95}$$

where

$$\sigma_s{}^2 = n(n - 1)(2n + 5)/18 \tag{96}$$

and d is taken to be normally distributed with zero mean and unit standard deviation. In employing this normal approximation it is as well to make a 'continuity' correction by replacing S by $S - 1$ if S is positive, and by $S + 1$ if S is negative.

The existence of tied rankings is a complication. When $n \leqslant 10$, tables exist for any number of tied pairs or triplets (Sillito, G. P., 1947, *Biometrika*, **34**, 36). When $n > 10$, we can employ the normal approximation above, but $\sigma_s{}^2$ should be modified. See *Rank Correlation Methods* by Kendall & Gibbons.

Appendix tables

The first five appendix tables have been mainly abridged from Tables I, III, IV, V and VI of Fisher and Yates' *Statistical Tables for Biological, Agricultural and Medical Research*, published by Oliver and Boyd Limited, Edinburgh, by permission of the authors and publishers. Some additional material has also been incorporated from Tables 12 and 18 of *Biometrika Tables for Statisticians*, Vol. I, by permission of the *Biometrika Trustees*. The sixth appendix table, originally by Professor John W. Tukey, has been taken directly from the Ciba–Geigy *Scientific Tables* (7th edn.), by permission of the author and publishers. Finally, the seventh appendix table adapted from Kendall, *Rank Correlation Methods* (4th edn.), 1970, has been reproduced by permission of the publishers, Charles Griffin and Company Ltd. of London and High Wycombe.

Appendix 1. *The normal distribution*

P	0.10	0.05	0.02	0.01	0.002	0.001
d	1.645	1.960	2.326	2.576	3.090	3.291

The table gives the percentage points most frequently required for significance tests and confidence limits based on a normal variable having zero mean and unit standard deviation (usually called d in the text). Thus, for any normal distribution, the probability of observing a departure from the mean of more than 1.960 standard deviations in *either* direction is 0.05 or 5 per cent.

Appendix 2. *'Student's' t-distribution*

Degrees of freedom	Value of P					
	0.10	0.05	0.02	0.01	0.002	0.001
1	6.314	12.71	31.82	63.66	318.3	636.6
2	2.920	4.303	6.965	9.925	22.33	31.60
3	2.353	3.182	4.541	5.841	10.21	12.92
4	2.132	2.776	3.747	4.604	7.173	8.610
5	2.015	2.571	3.365	4.032	5.893	6.869
6	1.943	2.447	3.143	3.707	5.208	5.959
7	1.895	2.365	2.998	3.499	4.785	5.408
8	1.860	2.306	2.896	3.355	4.501	5.041
9	1.833	2.262	2.821	3.250	4.297	4.781
10	1.812	2.228	2.764	3.169	4.144	4.587
11	1.796	2.201	2.718	3.106	4.025	4.437
12	1.782	2.179	2.681	3.055	3.930	4.318
13	1.771	2.160	2.650	3.012	3.852	4.221
14	1.761	2.145	2.624	2.977	3.787	4.140
15	1.753	2.131	2.602	2.947	3.733	4.073
16	1.746	2.120	2.583	2.921	3.686	4.015
17	1.740	2.110	2.567	2.898	3.646	3.965
18	1.734	2.101	2.552	2.878	3.610	3.922
19	1.729	2.093	2.539	2.861	3.579	3.883
20	1.725	2.086	2.528	2.845	3.552	3.850
21	1.721	2.080	2.518	2.831	3.527	3.819
22	1.717	2.074	2.508	2.819	3.505	3.792
23	1.714	2.069	2.500	2.807	3.485	3.767
24	1.711	2.064	2.492	2.797	3.467	3.745
25	1.708	2.060	2.485	2.787	3.450	3.725
26	1.706	2.056	2.479	2.779	3.435	3.707
27	1.703	2.052	2.473	2.771	3.421	3.690
28	1.701	2.048	2.467	2.763	3.408	3.674
29	1.699	2.045	2.462	2.756	3.396	3.659
30	1.697	2.042	2.457	2.750	3.385	3.646

The table gives the percentage points most frequently required for significance tests and confidence limits based on 'Student's' t-distribution. Thus, the probability of observing a value of t, with 10 degrees of freedom, greater in *absolute value* than 3.169 (i.e. < -3.169 or $> +3.169$) is exactly 0.01 or 1 per cent.

Appendix 3. *The χ^2 distribution*

Degrees of freedom	Value of P					
	0.99	0.95	0.05	0.10	0.01	0.001
1	0.000157	0.00393	3.841	2.706	6.635	10.83
2	0.0201	0.103	5.991	4.605	9.210	13.82
3	0.115	0.352	7.815	6.251	11.34	16.27
4	0.297	0.711	9.488	7.779	13.28	18.47
5	0.554	1.145	11.07	9.236	15.09	20.51
6	0.872	1.635	12.59	10.64	16.81	22.46
7	1.239	2.167	14.07	12.02	18.48	24.32
8	1.646	2.733	15.51	13.36	20.09	26.13
9	2.088	3.325	16.92	14.68	21.67	27.88
10	2.558	3.940	18.31	15.99	23.21	29.59
11	3.053	4.575	19.68	17.27	24.72	31.26
12	3.571	5.226	21.03	18.55	26.22	32.91
13	4.107	5.892	22.36	19.81	27.69	34.53
14	4.660	6.571	23.68	21.06	29.14	36.12
15	5.229	7.261	25.00	22.31	30.58	37.70
16	5.812	7.962	26.30	23.54	32.00	39.25
17	6.408	8.672	27.59	24.77	33.41	40.79
18	7.015	9.390	28.87	25.99	34.81	42.31
19	7.633	10.12	30.14	27.20	36.19	43.82
20	8.260	10.85	31.41	28.41	37.57	45.31
21	8.897	11.59	32.67	29.62	38.93	46.80
22	9.542	12.34	33.92	30.81	40.29	48.27
23	10.20	13.09	35.17	32.01	41.64	49.73
24	10.86	13.85	36.42	33.20	42.98	51.18
25	11.52	14.61	37.65	34.38	44.31	52.62
26	12.20	15.38	38.89	35.56	45.64	54.05
27	12.88	16.15	40.11	36.74	46.96	55.48
28	13.56	16.93	41.34	37.92	48.28	56.89
29	14.26	17.71	42.56	39.09	49.59	58.30
30	14.95	18.49	43.77	40.26	50.89	59.70

The table gives the percentage points most frequently required for significance tests based on χ^2. Thus, the probability of observing a χ^2 with 5 degrees of freedom *greater* in value than 11.07 is 0.05 or 5 per cent. Again, the probability of observing a χ^2 with 5 degrees of freedom *smaller* in value than 0.554 is $1 - 0.99 = 0.01$ or 1 per cent.

Appendix 4. *The correlation coefficient*

Degrees of freedom	Value of P				
	0.10	0.05	0.02	0.01	0.001
1	0.9877	0.99692	0.99951	0.99988	0.9999988
2	0.9000	0.9500	0.9800	0.9900	0.9990
3	0.805	0.878	0.9343	0.9587	0.9911
4	0.729	0.811	0.882	0.9172	0.9741
5	0.669	0.754	0.833	0.875	0.9509
6	0.621	0.707	0.789	0.834	0.9249
7	0.582	0.666	0.750	0.798	0.898
8	0.549	0.632	0.715	0.765	0.872
9	0.521	0.602	0.685	0.735	0.847
10	0.497	0.576	0.658	0.708	0.823
11	0.476	0.553	0.634	0.684	0.801
12	0.457	0.532	0.612	0.661	0.780
13	0.441	0.514	0.592	0.641	0.760
14	0.426	0.497	0.574	0.623	0.742
15	0.412	0.482	0.558	0.606	0.725
16	0.400	0.468	0.543	0.590	0.708
17	0.389	0.456	0.529	0.575	0.693
18	0.378	0.444	0.516	0.561	0.679
19	0.369	0.433	0.503	0.549	0.665
20	0.360	0.423	0.492	0.537	0.652
25	0.323	0.381	0.445	0.487	0.597
30	0.296	0.349	0.409	0.449	0.554
35	0.275	0.325	0.381	0.418	0.519
40	0.257	0.304	0.358	0.393	0.490
45	0.243	0.288	0.338	0.372	0.465
50	0.231	0.273	0.322	0.354	0.443
60	0.211	0.250	0.295	0.325	0.408
70	0.195	0.232	0.274	0.302	0.380
80	0.183	0.217	0.257	0.283	0.357
90	0.173	0.205	0.242	0.267	0.338
100	0.164	0.195	0.230	0.254	0.321

The table gives percentage points for the distribution of the estimated correlation coefficient r when the true value ρ is zero. Thus, when there are 10 degrees of freedom (i.e. in samples of 12) the probability of observing an r greater in *absolute value* than 0.576 (i.e. < -0.576 or $> +0.576$) is 0.05 or 5 per cent.

Appendix 5. *5 per cent points of variance-ratio (F) distribution*

f_2 \ f_1	1	2	3	4	5	6	7	8	9	10	12	15	20	30	∞
1	161.4	199.5	215.7	224.6	230.2	234.0	236.8	238.9	240.5	241.9	243.9	245.9	248.0	250.1	254.3
2	18.51	19.00	19.16	19.25	19.30	19.33	19.35	19.37	19.38	19.40	19.41	19.43	19.45	19.46	19.50
3	10.13	9.55	9.28	9.12	9.01	8.94	8.89	8.85	8.81	8.79	8.74	8.70	8.66	8.62	8.53
4	7.71	6.94	6.59	6.39	6.26	6.16	6.09	6.04	6.00	5.96	5.91	5.86	5.80	5.75	5.63
5	6.61	5.79	5.41	5.19	5.05	4.95	4.88	4.82	4.77	4.74	4.68	4.62	4.56	4.50	4.36
6	5.99	5.14	4.76	4.53	4.39	4.28	4.21	4.15	4.10	4.06	4.00	3.94	3.87	3.81	3.67
7	5.59	4.74	4.35	4.12	3.97	3.87	3.79	3.73	3.68	3.64	3.57	3.51	3.44	3.38	3.23
8	5.32	4.46	4.07	3.84	3.69	3.58	3.50	3.44	3.39	3.35	3.28	3.22	3.15	3.08	2.93
9	5.12	4.26	3.86	3.63	3.48	3.37	3.29	3.23	3.18	3.14	3.07	3.01	2.94	2.86	2.71
10	4.96	4.10	3.71	3.48	3.33	3.22	3.14	3.07	3.02	2.98	2.91	2.85	2.77	2.70	2.54
11	4.84	3.98	3.59	3.36	3.20	3.09	3.01	2.95	2.90	2.85	2.79	2.72	2.65	2.57	2.40
12	4.75	3.89	3.49	3.26	3.11	3.00	2.91	2.85	2.80	2.75	2.69	2.62	2.54	2.47	2.30
13	4.67	3.81	3.41	3.18	3.03	2.92	2.83	2.77	2.71	2.67	2.60	2.53	2.46	2.38	2.21
14	4.60	3.74	3.34	3.11	2.96	2.85	2.76	2.70	2.65	2.60	2.53	2.46	2.39	2.31	2.13
15	4.54	3.68	3.29	3.06	2.90	2.79	2.71	2.64	2.59	2.54	2.48	2.40	2.33	2.25	2.07
16	4.49	3.63	3.24	3.01	2.85	2.74	2.66	2.59	2.54	2.49	2.42	2.35	2.28	2.19	2.01
17	4.45	3.59	3.20	2.96	2.81	2.70	2.61	2.55	2.49	2.45	2.38	2.31	2.23	2.15	1.96
18	4.41	3.55	3.16	2.93	2.77	2.66	2.58	2.51	2.46	2.41	2.34	2.27	2.19	2.11	1.92
19	4.38	3.52	3.13	2.90	2.74	2.63	2.54	2.48	2.42	2.38	2.31	2.23	2.16	2.07	1.88
20	4.35	3.49	3.10	2.87	2.71	2.60	2.51	2.45	2.39	2.35	2.28	2.20	2.12	2.04	1.84

21	4.32	3.47	3.07	2.84	2.68	2.57	2.49	2.42	2.37	2.32	2.25	2.18	2.10	2.01	1.81
22	4.30	3.44	3.05	2.82	2.66	2.55	2.46	2.40	2.34	2.30	2.23	2.15	2.07	1.98	1.78
23	4.28	3.42	3.03	2.80	2.64	2.53	2.44	2.37	2.32	2.27	2.20	2.13	2.05	1.96	1.76
24	4.26	3.40	3.01	2.78	2.62	2.51	2.42	2.36	2.30	2.25	2.18	2.11	2.03	1.94	1.73
25	4.24	3.39	2.99	2.76	2.60	2.49	2.40	2.34	2.28	2.24	2.16	2.09	2.01	1.92	1.71
26	4.23	3.37	2.98	2.74	2.59	2.47	2.39	2.32	2.27	2.22	2.15	2.07	1.99	1.90	1.69
27	4.21	3.35	2.96	2.73	2.57	2.46	2.37	2.31	2.25	2.20	2.13	2.06	1.97	1.88	1.67
28	4.20	3.34	2.95	2.71	2.56	2.45	2.36	2.29	2.24	2.19	2.12	2.04	1.96	1.87	1.65
29	4.18	3.33	2.93	2.70	2.55	2.43	2.35	2.28	2.22	2.18	2.10	2.03	1.94	1.85	1.64
30	4.17	3.32	2.92	2.69	2.53	2.42	2.33	2.27	2.21	2.16	2.09	2.01	1.93	1.84	1.62
40	4.08	3.23	2.84	2.61	2.45	2.34	2.25	2.18	2.12	2.08	2.00	1.92	1.84	1.74	1.51
60	4.00	3.15	2.76	2.53	2.37	2.25	2.17	2.10	2.04	1.99	1.92	1.84	1.75	1.65	1.39
120	3.92	3.07	2.68	2.45	2.29	2.17	2.09	2.02	1.96	1.91	1.83	1.75	1.66	1.55	1.25
∞	3.84	3.00	2.60	2.37	2.21	2.10	2.01	1.94	1.88	1.83	1.75	1.67	1.57	1.46	1.00

The table gives the 5 per cent points of the distribution of the variance-ratio. $F = s_1^2/s_2^2$, where the numerator and denominator have f_1 and f_2 degrees of freedom respectively. Thus, if $f_1 = 7$ and $f_2 = 15$, the probability that the observed value of F is *greater* than 2.71 is exactly 0.05 or 5 per cent.

Appendix 5 (*continued*). *1 per cent points of variance-ratio* (F) *distribution*

f_2 \ f_1	1	2	3	4	5	6	7	8	9	10	12	15	20	30	∞
1	4052	4999	5403	5625	5764	5859	5928	5982	6022	6056	6106	6157	6209	6261	6366
2	98.50	99.00	99.17	99.25	99.30	99.33	99.36	99.37	99.39	99.40	99.42	99.43	99.45	99.47	99.50
3	34.12	30.82	29.46	28.71	28.24	27.91	27.67	27.49	27.35	27.23	27.05	26.87	26.69	26.50	26.13
4	21.20	18.00	16.69	15.98	15.52	15.21	14.98	14.80	14.66	14.55	14.37	14.20	14.02	13.84	13.46
5	16.26	13.27	12.06	11.39	10.97	10.67	10.46	10.29	10.16	10.05	9.89	9.72	9.55	9.38	9.02
6	13.75	10.92	9.78	9.15	8.75	8.47	8.26	8.10	7.98	7.87	7.72	7.56	7.40	7.23	6.88
7	12.25	9.55	8.45	7.85	7.46	7.19	6.99	6.84	6.72	6.62	6.47	6.31	6.16	5.99	5.65
8	11.26	8.65	7.59	7.01	6.63	6.37	6.18	6.03	5.91	5.81	5.67	5.52	5.36	5.20	4.86
9	10.56	8.02	6.99	6.42	6.06	5.80	5.61	5.47	5.35	5.26	5.11	4.96	4.81	4.65	4.31
10	10.04	7.56	6.55	5.99	5.64	5.39	5.20	5.06	4.94	4.85	4.71	4.56	4.41	4.25	3.91
11	9.65	7.21	6.22	5.67	5.32	5.07	4.89	4.74	4.63	4.54	4.40	4.25	4.10	3.94	3.60
12	9.33	6.93	5.95	5.41	5.06	4.82	4.64	4.50	4.39	4.30	4.16	4.01	3.86	3.70	3.36
13	9.07	6.70	5.74	5.21	4.86	4.62	4.44	4.30	4.19	4.10	3.96	3.82	3.66	3.51	3.17
14	8.86	6.51	5.56	5.04	4.69	4.46	4.28	4.14	4.03	3.94	3.80	3.66	3.51	3.35	3.00
15	8.68	6.36	5.42	4.89	4.56	4.32	4.14	4.00	3.89	3.80	3.67	3.52	3.37	3.21	2.87
16	8.53	6.23	5.29	4.77	4.44	4.20	4.03	3.89	3.78	3.69	3.55	3.41	3.26	3.10	2.75
17	8.40	6.11	5.18	4.67	4.34	4.10	3.93	3.79	3.68	3.59	3.46	3.31	3.16	3.00	2.65
18	8.29	6.01	5.09	4.58	4.25	4.01	3.84	3.71	3.60	3.51	3.37	3.23	3.08	2.92	2.57
19	8.18	5.93	5.01	4.50	4.17	3.94	3.77	3.63	3.52	3.43	3.30	3.15	3.00	2.84	2.49
20	8.10	5.85	4.94	4.43	4.10	3.87	3.70	3.56	3.46	3.37	3.23	3.09	2.94	2.78	2.42

21	8.02	5.78	4.87	4.37	4.04	3.81	3.64	3.51	3.40	3.31	3.17	3.03	2.88	2.72	2.36
22	7.95	5.72	4.82	4.31	3.99	3.76	3.59	3.45	3.35	3.26	3.12	2.98	2.83	2.67	2.31
23	7.88	5.66	4.76	4.26	3.94	3.71	3.54	3.41	3.30	3.21	3.07	2.93	2.78	2.62	2.26
24	7.82	5.61	4.72	4.22	3.90	3.67	3.50	3.36	3.26	3.17	3.03	2.89	2.74	2.58	2.21
25	7.77	5.57	4.68	4.18	3.85	3.63	3.46	3.32	3.22	3.13	2.99	2.85	2.70	2.54	2.17
26	7.72	5.53	4.64	4.14	3.82	3.59	3.42	3.29	3.18	3.09	2.96	2.81	2.66	2.50	2.13
27	7.68	5.49	4.60	4.11	3.78	3.56	3.39	3.26	3.15	3.06	2.93	2.78	2.63	2.47	2.10
28	7.64	5.45	4.57	4.07	3.75	3.53	3.36	3.23	3.12	3.03	2.90	2.75	2.60	2.44	2.06
29	7.60	5.42	4.54	4.04	3.73	3.50	3.33	3.20	3.09	3.00	2.87	2.73	2.57	2.41	2.03
30	7.56	5.39	4.51	4.02	3.70	3.47	3.30	3.17	3.07	2.98	2.84	2.70	2.55	2.39	2.01
40	7.31	5.18	4.31	3.83	3.51	3.29	3.12	2.99	2.89	2.80	2.66	2.52	2.37	2.20	1.80
60	7.08	4.98	4.13	3.65	3.34	3.12	2.95	2.82	2.72	2.63	2.50	2.35	2.20	2.03	1.60
120	6.85	4.79	3.95	3.48	3.17	2.96	2.79	2.66	2.56	2.47	2.34	2.19	2.03	1.86	1.38
∞	6.63	4.61	3.78	3.32	3.02	2.80	2.64	2.51	2.41	2.32	2.18	2.04	1.88	1.70	1.00

The table gives the 1 per cent points of the distribution of the variance-ratio, $F = s_1^2/s_2^2$, where the numerator and denominator have f_1 and f_2 degrees of freedom respectively. Thus, if $f_1 = 7$ and $f_2 = 15$, the probability that the observed value of F is *greater* than 4.14 is exactly 0.01 or 1 per cent.

Appendix 6. *Wilcoxon's test for paired comparisons*

Sample size n	Value of P			
	0.10	0.05	0.02	0.01
5	0, 15	–	–	–
6	2, 19	0, 21	–	–
7	3, 25	2, 26	0, 28	–
8	5, 31	3, 33	1, 35	0, 36
9	8, 37	5, 40	3, 42	1, 44
10	10, 45	8, 47	5, 50	3, 52
11	13, 53	10, 56	7, 59	5, 61
12	17, 61	13, 65	9, 69	7, 71
13	21, 70	17, 74	12, 79	9, 82
14	25, 80	21, 84	15, 90	12, 93
15	30, 90	25, 95	19, 101	15, 105
16	35, 101	29, 107	23, 113	19, 117
17	41, 112	34, 199	28, 125	23, 130
18	47, 124	40, 131	32, 139	27, 144
19	53, 137	46, 144	37, 153	32, 158
20	60, 150	52, 158	43, 167	37, 173
21	67, 164	58, 173	49, 182	42, 189
22	75, 178	66, 187	55, 198	48, 205
23	83, 193	73, 203	62, 214	54, 222
24	91, 209	81, 219	69, 231	61, 239
25	100, 225	89, 236	76, 249	68, 257

The table gives the extreme values for T in either direction that must be *attained or exceeded* for significance to be achieved at the level indicated.

Appendix 7. *Kendall's rank correlation coefficient* τ

Sample size n	Value of P			
	0.10	0.05	0.02	0.01
4	−6 , 6	–	–	–
5	−8 , 8	−10, 10	−10, 10	–
6	−11, 11	−13, 13	−13, 13	−15, 15
7	−13, 13	−15, 15	−17, 17	−19, 19
8	−16, 16	−18, 18	−20, 20	−22, 22
9	−18, 18	−20, 20	−24, 24	−26, 26
10	−21, 21	−23, 23	−27, 27	−29, 29

The table gives the extreme values of S in either direction that must be *attained or exceeded* for significance to be achieved at the level indicated.

Index